Hodder Gibson

Scottish Examination Materia

HIGHER BIOLOGY

Multiple Choice & Matching

SECOND EDITION

Team Co-ordinator: James Torrance

Writing Team: James Torrance
James Fullarton
Clare Marsh
James Simms
Caroline Stevenson

Diagrams by James Torrance

Orders: please contact Bookpoint Ltd, 130 Milton Park, Abingdon, Oxon OX14 4SB. Telephone: (44) 01235 827720. Fax: (44) 01235 400454. Lines are open 9.00–5.00, Monday to Saturday, with a 24-hour message answering service. Visit our website at www.hoddereducation.co.uk. Hodder Gibson can be contacted direct on: Tel: 0141 848 1609; Fax: 0141 889 6315; email: hoddergibson@hodder.co.uk

First published in 1999 by
Hodder Gibson, an imprint of Hodder Education,
An Hachette Livre UK Company,
2a Christie Street
Paisley PA1 1NB

This second edition first published (2008)

Impression number 5 4 3 2 1
Year 2012 2011 2010 2009 2008

Cover photo © Steve Bloom Images/Alamy
Illustrations by James Torrance
Typeset in 11/14 Sabon by Macmillan Publishing Solutions (www.macmillansolutions.com)
Printed and bound in Great Britain by Martins the Printers, Berwick-upon-Tweed

A catalogue record for this title is available from the British Library

ISBN-13: 978 0340 973 042

Contents

Preface

This book has been written specifically to complement the textbook *Higher Biology 2nd Edition*. It is intended to act as a valuable resource to pupils and teachers by providing a set of matching exercises and a comprehensive bank of multiple choice items, the content of which adheres closely to the SQA Higher Still syllabus for Higher Grade Biology.

Each test corresponds to part of a syllabus sub-topic. The matching exercises enable pupils to gradually construct a glossary of terms essential to the course. The multiple choice components contain a variety of types of item, many testing *knowledge* and *understanding*, some testing *problem-solving* skills and others testing *practical abilities*. These allow pupils to practise extensively in preparation for the examination. The book concludes with two 30-item specimen examinations in the style of the multiple choice section of the externally assessed higher examination paper.

1

Cell variety in relation to function

Match the terms in list X with their descriptions in list Y.

list X		list Y	
1	cell wall	**a**	term used to describe an organism which consists of one cell
2	chloroplast	**b**	term used to describe an organism which consists of more than one cell
3	cheek epithelium	**c**	site of digestion in a unicellular animal
4	cilia	**d**	plant tissue responsible for support and water transport
5	ciliated epithelium	**e**	control centre of a cell containing genetic material
6	contractile vacuole	**f**	green structure responsible for photosynthesis in a plant cell
7	epidermis	**g**	group of cells specialised to perform a particular function
8	food vacuole	**h**	hair-like structures used by some unicellular animals for locomotion
9	mesophyll	**i**	structure which expels excess water from a unicellular animal's body
10	motor neurone	**j**	protective tissue consisting of irregularly shaped cells which are replaced constantly
11	multicellular	**k**	protective tissue which secretes mucus and bears cilia
12	nucleus	**l**	type of nerve cell which transmits a nerve impulse to an effector
13	phloem	**m**	protective tissue on the outside of plant organs
14	tissue	**n**	structure which provides a plant cell with support and protection
15	unicellular	**o**	photosynthetic tissue present in a leaf
16	xylem	**p**	plant tissue responsible for translocation of sugar

Choose the ONE correct answer to each of the following multiple choice questions.

1 Which of the following is the basic unit of life (i.e. the smallest unit that can lead an independent existence)?
A molecule **B** tissue **C** cell **D** nucleus

Items 2 and 3 refer to the following diagram of a unicellular organism called *Chlamydomonas* which is able to swim about in its natural habitat (stagnant water).

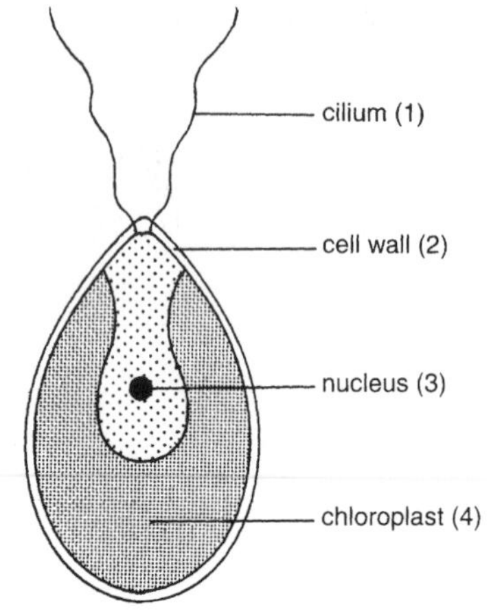

2 *Chlamydomonas* is
A an animal
B a bacterium
C a fungus
D a green plant

3 In the following table, which set of numbered structures corresponds correctly to the functions that they perform?

	function			
	control of cell activities	*movement*	*photosynthesis*	*protection*
A	1	3	4	2
B	3	1	2	4
C	4	2	3	1
D	3	1	4	2

Items 4 and 5 refer to the following diagram of the organism *Euglena*.

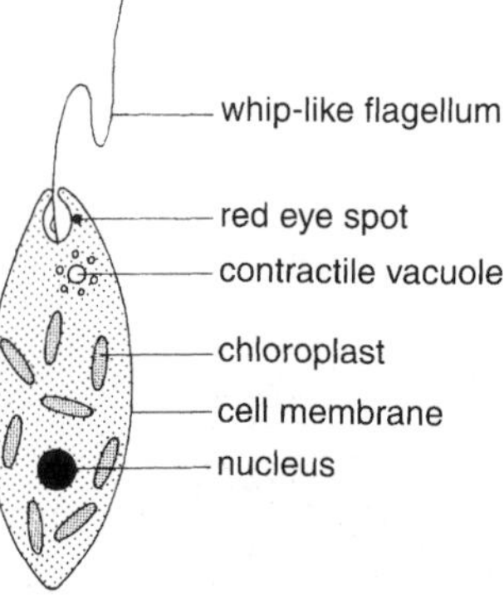

4 *Euglena* could be classified as an animal due to the
A absence of a cell wall. **B** presence of a nucleus.
C absence of motile cilia. **D** presence of a cell membrane.

5 *Euglena* could be classified as a plant due to the presence of a
A cell membrane. **B** chloroplast.
C contractile vacuole. **D** flagellum.

6 The function of goblet cells present in ciliated epithelium in the human trachea is to

A release carbon dioxide.
B sweep microbes away.
C secrete mucus.
D absorb oxygen.

Items 7, 8 and 9 refer to the accompanying diagrams of human cells (not drawn to scale).

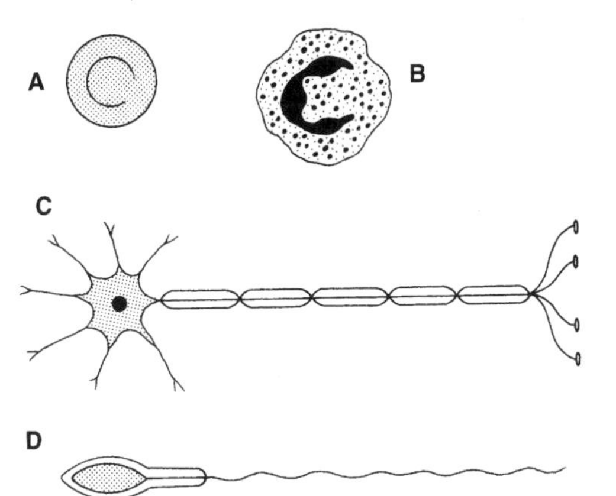

7 Which cell's structure is suited to the function of transmitting nerve impulses?

8 Which cell's structure is suited to the function of oxygen uptake?

9 Which cell's structure is suited to digesting and destroying bacteria?

10 Epithelium from the trachea and the internal lining of the cheek BOTH possess

A squamous cells.
B columnar cells.
C ciliated cells.
D goblet cells.

11 Which of the following tubes present in the human body is lined with ciliated epithelium?

A oviduct
B artery
C intestine
D vein

Items **12, 13, 14** and **15** refer to the following diagram of a transverse section through part of a leaf.

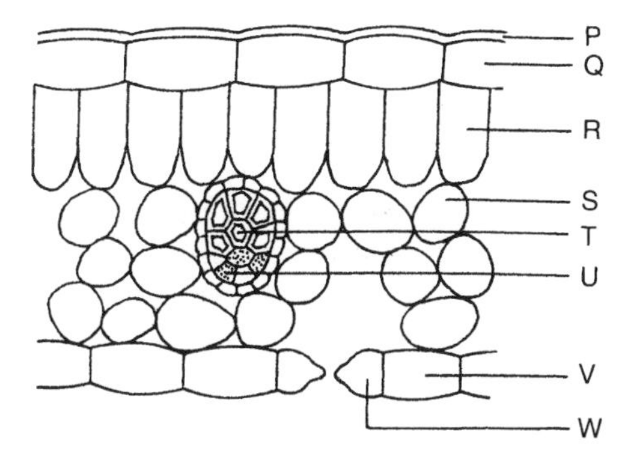

12 Which of the following are BOTH composed of cells whose structure suits their function of protection?

A P and Q **B** Q and V
C P and V **D** P and W

13 Photosynthesis occurs in

A Q, R and S **B** R, S and W
C P, Q and R **D** R, S and V

14 Which cells are structurally adapted to suit their function of controlling stomatal size?

A T **B** U **C** V **D** W

15 Which tissue contains cells structurally adapted to translocate soluble carbohydrates to other parts of the plant?

A R **B** S **C** T **D** U

Items **16, 17, 18** and **19** refer to the accompanying diagram (see opposite page) which shows a transverse section of a plant's stem and close-ups of three of its tissues.

16 Which of the answers that follow correctly describes tissue Q?

17 Which of the answers correctly describes tissue R?

18 Which of the answers correctly describes tissue S?

A It is found at location 1 and it protects the plant.
B It is found at location 2 and it is the site of sugar transport.
C It is found at location 3 and it supports the plant.
D It is found in location 4 and it is the site of water transport.

19 The name of tissue R is

A cortex. **B** xylem. **C** phloem. **D** epidermis.

20 The diagram shows a type of plant cell.
Compared with this cell, a mature xylem vessel would possess

A 1 only.
B 1 and 2 only.
C 1, 2 and 3 only.
D 1, 2, 3 and 4.

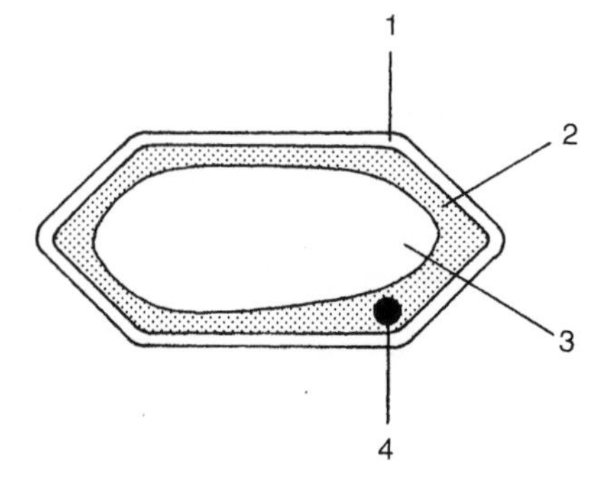

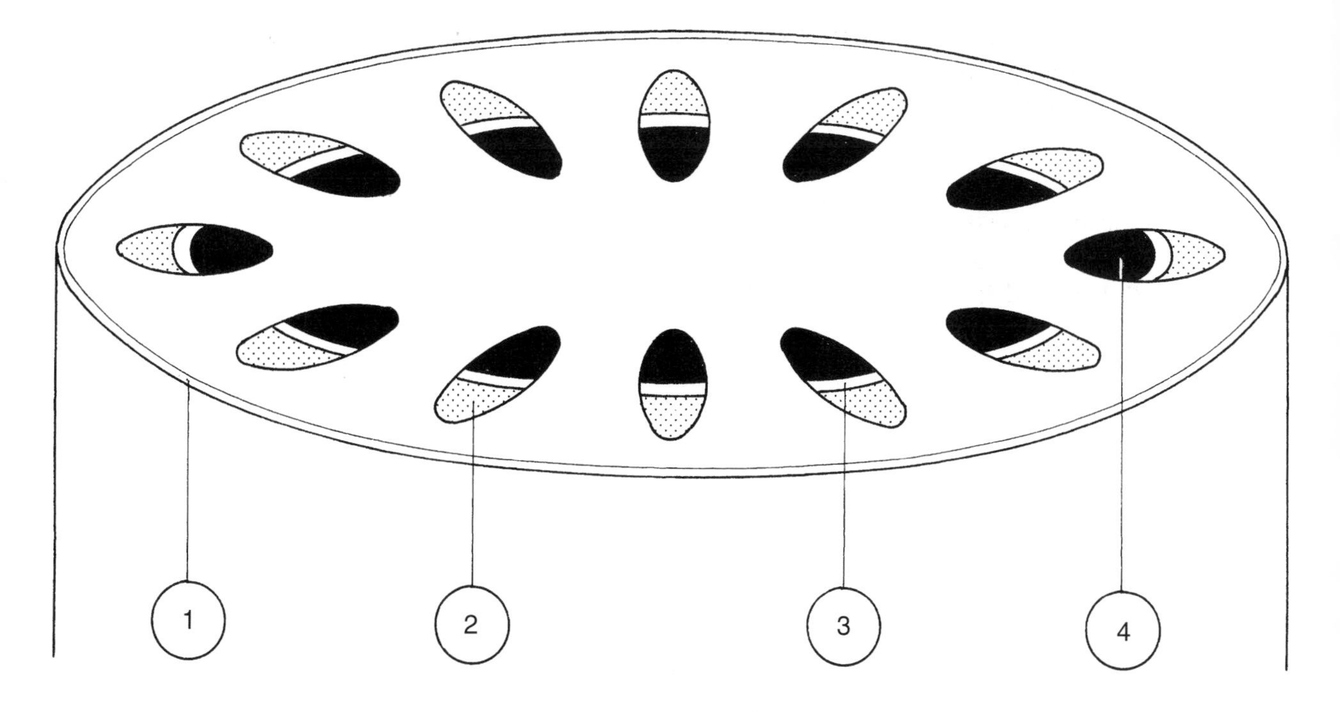

Tissue Q

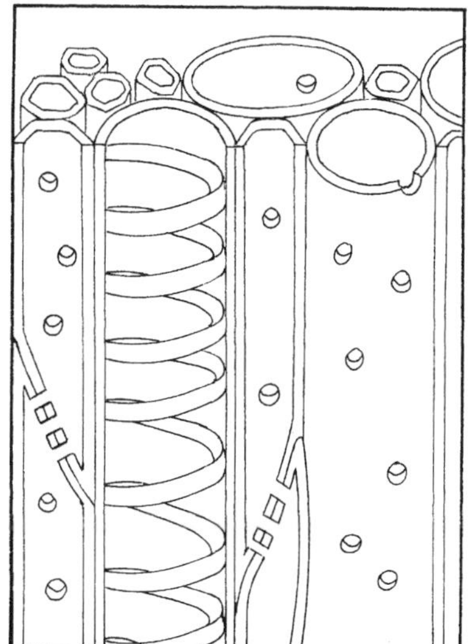

Tissue R

Tissue S

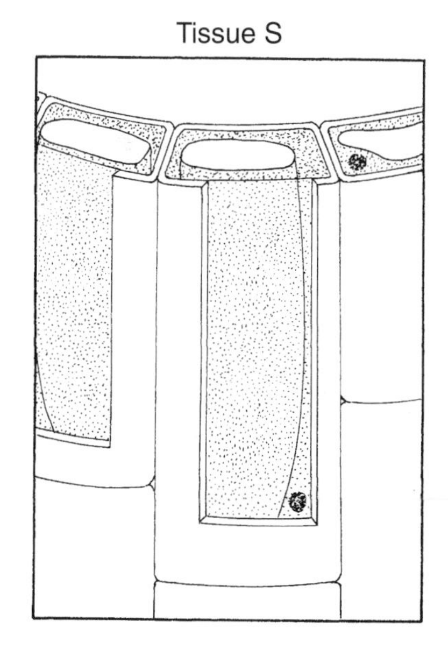

2 Absorption and secretion of materials

Match the terms in list X with their descriptions in list Y.

list X	list Y
1 active transport	**a** structure that allows rapid passage through it of small molecules (e.g. water) but not larger molecules
2 concentration gradient	**b** term used to describe two solutions that are equal in water concentration
3 diffusion	**c** the difference in concentration that exists between two regions resulting in diffusion
4 flaccid	**d** term used to describe a structure that allows rapid passage through it of all molecules in solution
5 freely permeable	**e** shrinkage of plant cell contents away from cell walls as a result of excessive water loss
6 hypertonic	**f** term used to describe plant tissue which has lost water by osmosis and become soft
7 hypotonic	**g** term used to describe a solution with a higher water concentration than a comparable solution
8 isotonic	**h** term used to describe a plant cell or tissue swollen with water taken in by osmosis
9 osmosis	**i** term used to describe a solution with a lower water concentration than a comparable solution
10 plasmolysis	**j** movement of molecules or ions from a region of higher concentration to a region of lower concentration of that type of molecule or ion
11 selectively permeable membrane	**k** net movement of water molecules from a region of higher water concentration to a region of lower water concentration through a selectively permeable membrane
12 turgid	**l** movement of molecules or ions through a plasma membrane from a region of lower concentration to a region of higher concentration of that type of molecule or ion

Choose the ONE correct answer to each of the following multiple choice questions.

1 The diagram opposite shows ways in which molecules may move into and out of a respiring animal cell. Which of these could be diffusion of carbon dioxide molecules?

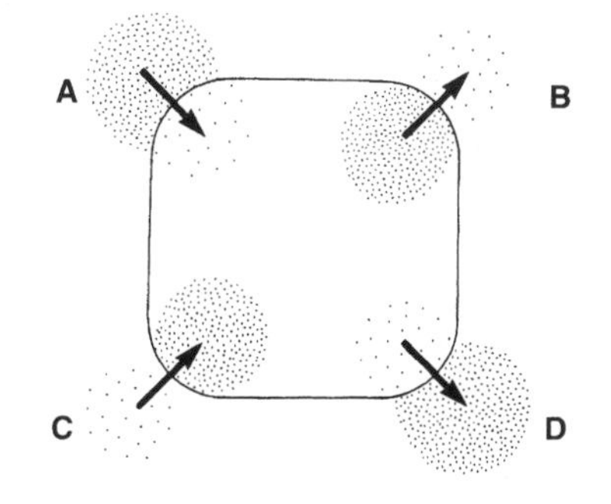

2 Osmosis is the passage through a selectively permeable membrane of

A water from a region of higher solute concentration to a region of lower solute concentration.
B solute from a region of lower water concentration to a region of higher water concentration.
C water from a region of lower solute concentration to a region of higher solute concentration.
D solute from a region of higher water concentration to a region of lower water concentration.

Items 3 and 4 refer to the experiment shown in the following diagram.

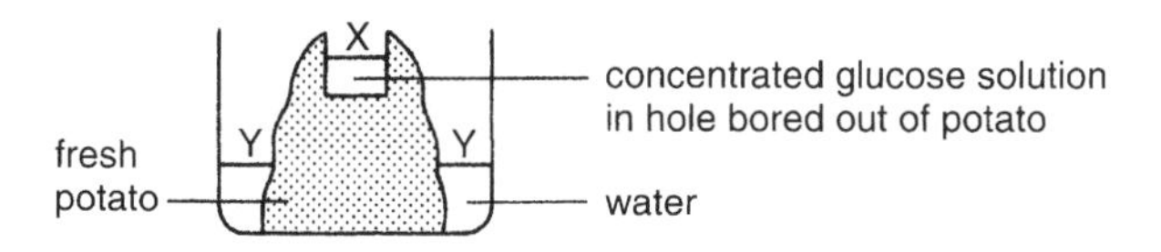

3 After a few days, which of the following will have occurred?

A a rise in level X and a drop in level Y
B a drop in level X and a drop in level Y
C a rise in level X and a rise in level Y
D a drop in level X and a rise in level Y

4 Which of the following diagrams shows a suitable control for the above experiment?

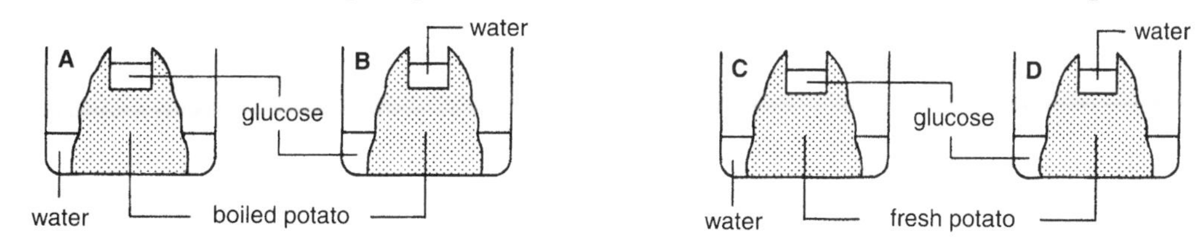

5 Which of the following are BOTH present in a cell wall?

A cellulose fibres and phospholipids
B phospholipids and plasma membranes
C plasma membranes and water-filled spaces
D water-filled spaces and cellulose fibres

6 The following diagram shows the fluid-mosaic model of the structure of a cell membrane.

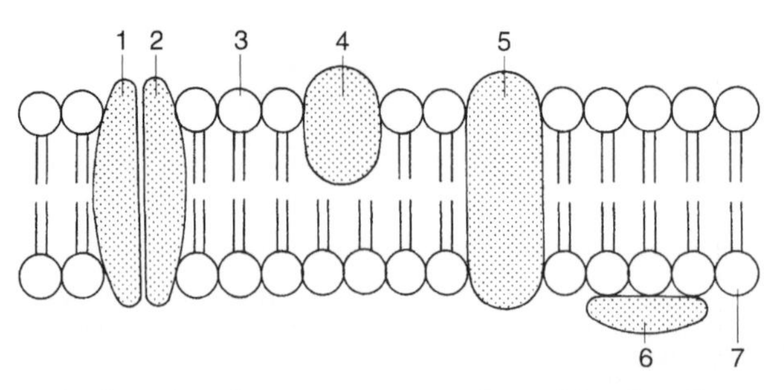

Which pair of structures numbered in the diagram are correctly identified in the following table?

	protein	*phospholipid*
A	1	7
B	2	4
C	3	7
D	5	6

7 Which of the following sucrose solutions has the highest water concentration?

A 1.1 molar **B** 0.8 molar

C 0.5 molar **D** 0.1 molar

8 The diagram shows the appearance of a plant cell immersed in a solution which is isotonic to the cell's sap.

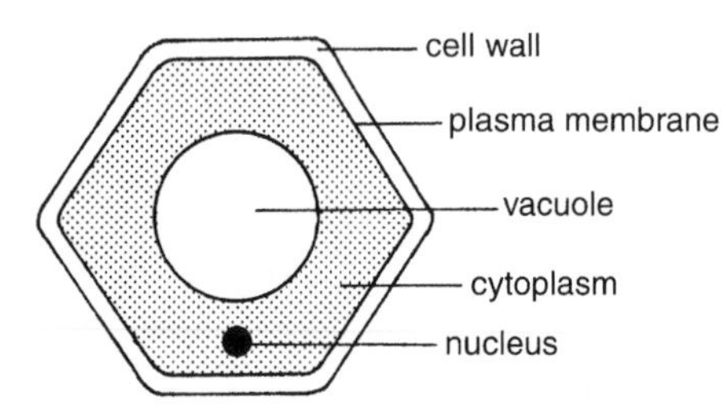

Which of the diagrams shown below most accurately represents the appearance of this cell after immersion in a hypertonic solution?

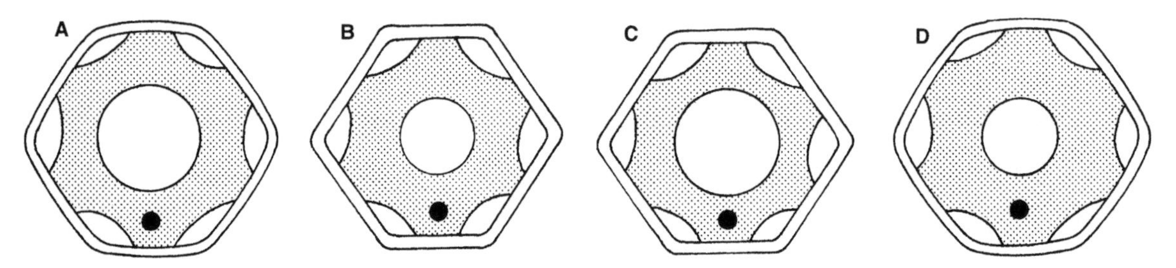

Items **9** and **10** refer to the following graph of the results from an experiment where each of the four potato cylinders was immersed in a different chemical for a time and then placed in water.

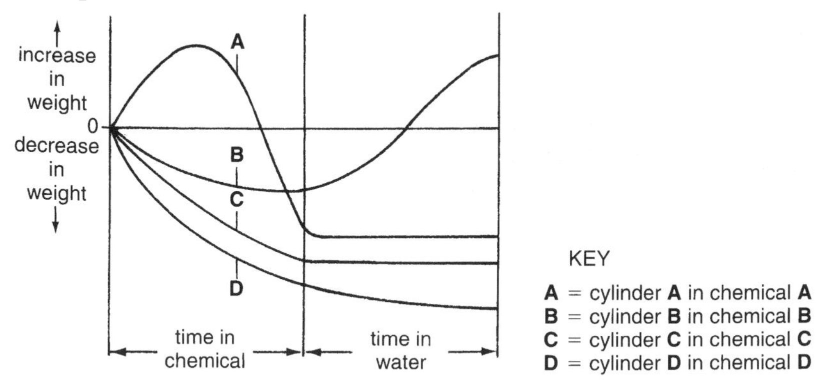

9 Which chemical was NOT toxic to the selectively permeable membranes of potato cells?

10 Deplasmolysis is the opposite process from plasmolysis. In which cylinder did deplasmolysis occur?

Items **11** and **12** refer to the following diagram of four plant cells.

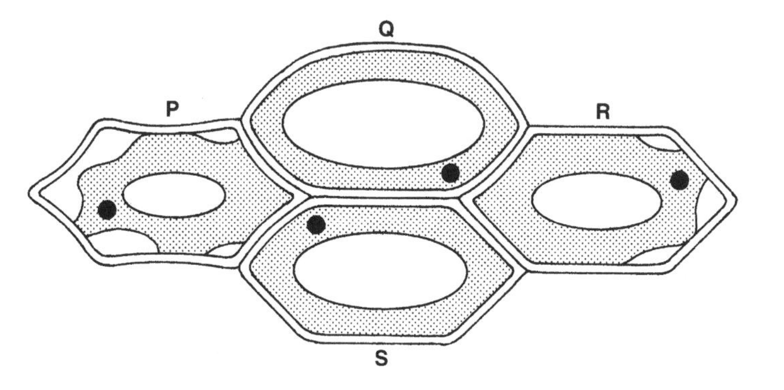

11 No wall pressure would exist in cells
A P and Q. **B** Q and S. **C** P and R. **D** R and S.

12 If the cells remained in contact as shown, then water would pass by osmosis from BOTH
A R to Q and Q to P. **B** Q to S and R to Q.
C P to Q and R to S. **D** Q to P and Q to R.

13 The diagram below shows the results of an analysis of cell sap from a marine plant and the surrounding sea water.

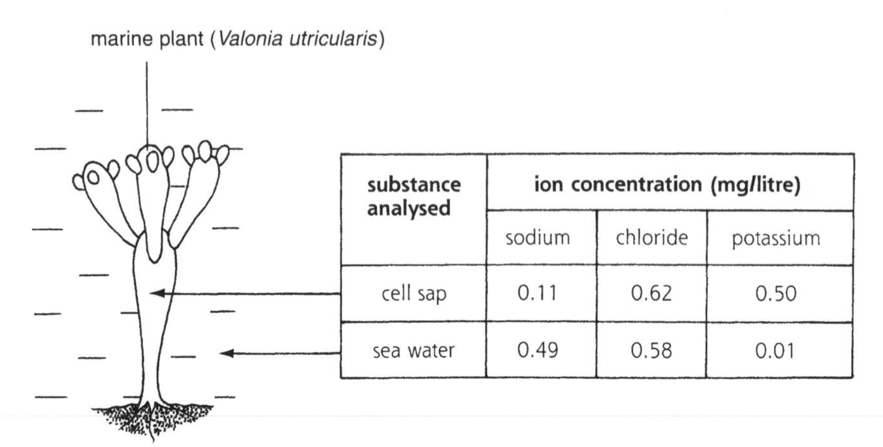

substance analysed	ion concentration (mg/litre)		
	sodium	chloride	potassium
cell sap	0.11	0.62	0.50
sea water	0.49	0.58	0.01

From these data it can be concluded that this plant

A accumulates all three types of ion from sea water.
B holds chloride ions at a concentration lower than sea water.
C selects and internally accumulates sodium ions only.
D can discriminate between sodium and potassium ions.

14 The table below shows the outcome of an investigation into the uptake of bromide ions by a plant.

time from start of experiment in minutes	*units of bromide ions taken up by plant tissue under the following conditions*		
	sugar absent, oxygen present	*sugar present, oxygen absent*	*sugar and oxygen present*
0	0	0	0
30	0	30	100
60	0	50	150
90	0	70	180
120	0	70	200

These results indicate that uptake of bromide ions

A is an active process requiring energy.
B occurs during aerobic respiration only.
C requires a temperature suitable for enzymes to act.
D stops in the absence of oxygen.

Items **15, 16** and **17** refer to the following diagram of the transport of two types of ion through a cell membrane. The carrier molecule is called the sodium/potassium pump since it exchanges one type of ion for the other.

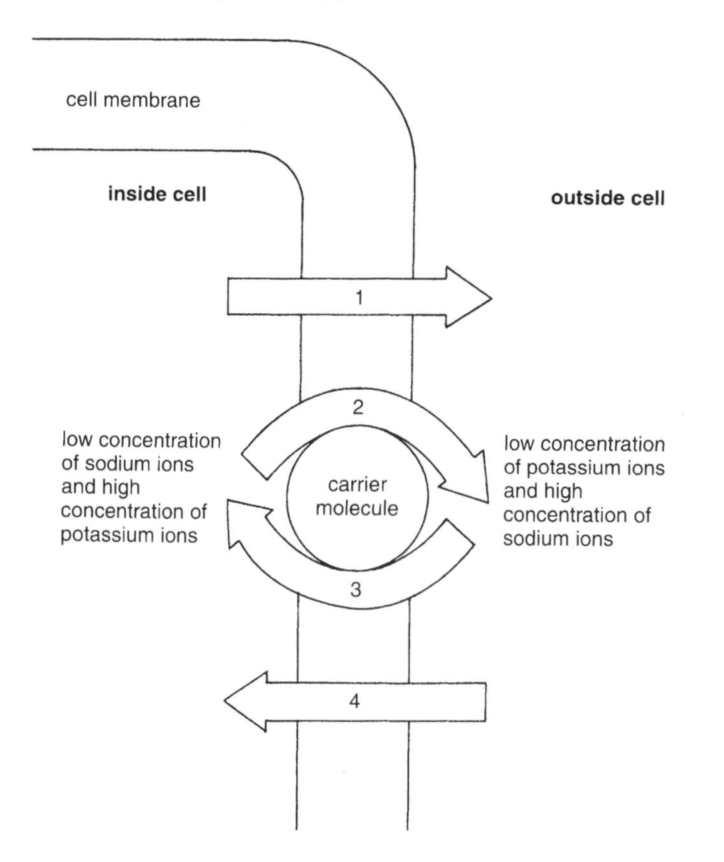

15 Active transport of sodium ions occurs at arrow
A 1. **B** 2. **C** 3. **D** 4.

16 Diffusion of potassium ions occurs at arrow
A 1. **B** 2. **C** 3. **D** 4.

17 Cyanide is a respiratory poison. Application of cyanide to the cell shown would affect and bring to a halt the processes indicated by arrows
A 1 and 4 only. **B** 3 and 4 only.
C 2 and 3 only. **D** 1, 2, 3 and 4.

Items **18** and **19** refer to the following graphs.

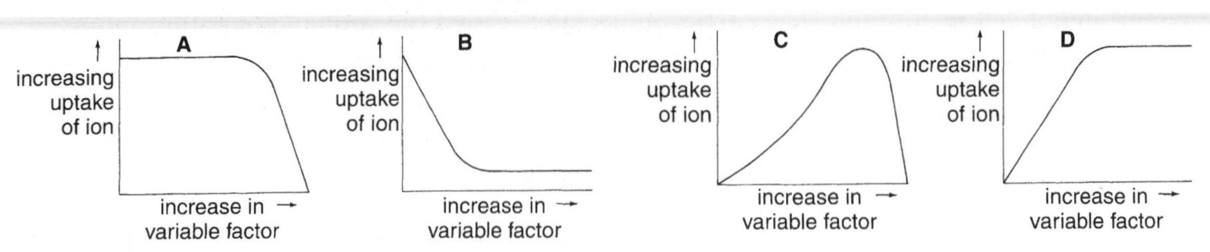

18 Which graph represents rate of ion uptake by a living cell in response to increasing oxygen concentration?

19 Which graph represents rate of ion uptake by a living cell in response to increasing temperature?

20 The following graph shows the changes in ionic concentrations of culture solutions in which barley roots were grown for two days.

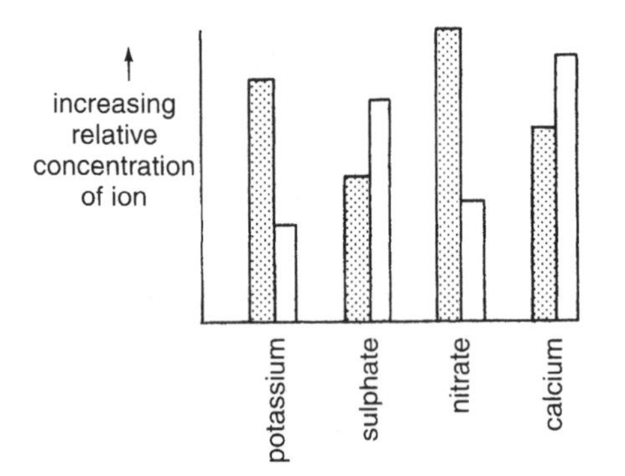

Which of the following ions were BOTH taken up by the plant?

A calcium and sulphate
B sulphate and potassium
C potassium and nitrate
D nitrate and calcium

3

ATP and energy release

Match the terms in list X with their descriptions in list Y.

list X

1 ADP
2 ATP
3 energy transfer
4 glucose
5 oxidation
6 phosphorylation
7 Pi
8 reduction
9 respiration

list Y

a process such as respiration where hydrogen is removed from a substrate
b process such as photosynthesis where hydrogen is added to a substrate
c low energy molecule composed of adenosine and two phosphate groups
d process by which ATP is regenerated from ADP and Pi
e biochemical pathway (aerobic or anaerobic) which releases energy
f respiratory substrate which provides energy for regeneration of ATP
g role played by ATP between energy-releasing and energy-consuming reactions
h high energy molecule composed of adenosine and three phosphate groups
i inorganic phosphate group needed to make ATP

Choose ONE correct answer to each of the following multiple choice questions.

1 Which of the following diagrams best represents the structure of a molecule of ATP (adenosine triphosphate)?

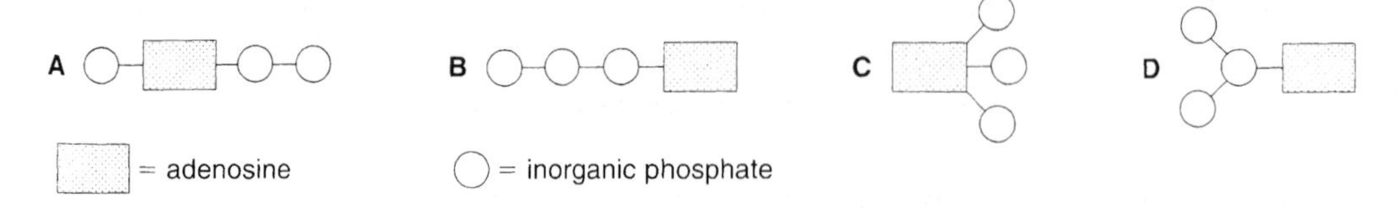

2 Which of the following equations represents the regeneration of ATP from its components?

A ADP + Pi $\xrightarrow{\text{energy taken in}}$ ATP

B ADP + Pi + Pi $\xrightarrow{\text{energy taken in}}$ ATP

C ADP + Pi $\xrightarrow{\text{energy released}}$ ATP

D ADP + Pi + Pi $\xrightarrow{\text{energy released}}$ ATP

3 The regeneration of ATP from its components is called

A oxidation. **B** metabolism.
C respiration. **D** phosphorylation.

4 As part of an investigation into the effect of different solutions on fresh muscle tissue, 12 drops of ATP were added to a strand of fresh muscle of initial length 50 mm. After a few minutes its length was found to be 42 mm.
Which of the following correctly summarises the experiment?

	% difference in length of muscle strand	*reason for change*
A	8	contraction of muscle fibres
B	8	relaxation of muscle fibres
C	16	contraction of muscle fibres
D	16	relaxation of muscle fibres

Items 5 and 6 refer to the following possible answers.

A removal of hydrogen ions from substrate and release of energy
B addition of hydrogen ions to substrate and consumption of energy
C removal of hydrogen ions from substrate and consumption of energy
D addition of hydrogen ions to substrate and release of energy

5 Which statement describes the process of oxidation?

6 Which statement describes the process of reduction?

Items 7 and 8 refer to the following diagram of tissue respiration.

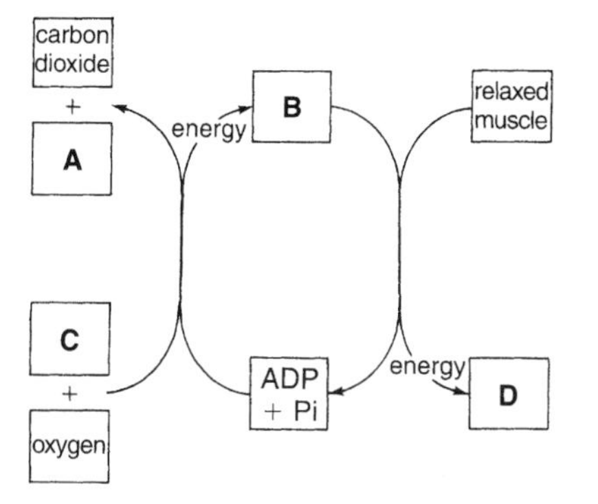

7 Which box represents the correct position of ATP in this scheme?

8 Which box represents the correct position of glucose in this scheme?

9 A muscle will contract in the presence of

A ADP at 60°C. **B** ATP at 60°C.
C ADP at 30°C. **D** ATP at 30°C.

10 Which of the following is an example of anabolism (the synthetic reactions of metabolism which require energy from ATP)?

A deamination of amino acids to form urea
B digestion of starch by enzymes
C formation of protein from amino acids
D oxidation of glucose during respiration

4 Chemistry of respiration

Match the terms in list X with their descriptions in list Y.

list X		**list Y**	
1	acetyl CoA	**a**	stage of respiratory pathway common to both aerobic and anaerobic respiration
2	ATP	**b**	series of hydrogen carriers located on the cristae of a mitochondrion
3	carbon dioxide	**c**	stage of respiratory pathway that occurs in the central matrix of a mitochondrion
4	central matrix	**d**	folded extension of the inner membrane of a mitochondrion
5	citric acid	**e**	organelle responsible for aerobic respiration
6	crista	**f**	the site in a cell where glycolysis occurs
7	cytochrome system	**g**	part of a mitochondrion containing enzymes needed for Krebs' cycle
8	cytoplasm	**h**	final hydrogen acceptor in aerobic respiration
9	ethanol	**i**	product of aerobic respiration when oxygen combines with hydrogen
10	glucose	**j**	high energy compound formed by phosphorylation using energy released during respiration
11	glycolysis	**k**	coenzyme which accepts hydrogen during aerobic respiration and passes it to the cytochrome system
12	Krebs' cycle	**l**	process by which high energy ATP is formed from low energy ADP + Pi
13	lactic acid	**m**	product of aerobic and anaerobic respiration in plant cells
14	mitochondrion	**n**	2-carbon compound formed from pyruvic acid and CoA in the presence of oxygen
15	NAD	**o**	2-carbon compound produced in plant cells during anaerobic respiration
16	oxygen	**p**	3-carbon compound produced in animal cells during anaerobic respiration
17	phosphorylation	**q**	3-carbon compound formed from glucose during glycolysis
18	pyruvic acid	**r**	6-carbon compound broken down by enzyme action in Krebs' cycle
19	water	**s**	6-carbon sugar which acts as a respiratory substrate in both aerobic and anaerobic respiration

Choose the ONE correct answer to each of the following multiple choice questions.

Items **1, 2, 3** and **4** refer to the following diagram which shows a simplified summary of aerobic respiration. (Each intermediate compound is represented by a box containing the number of carbon atoms present in one molecule of the compound.)

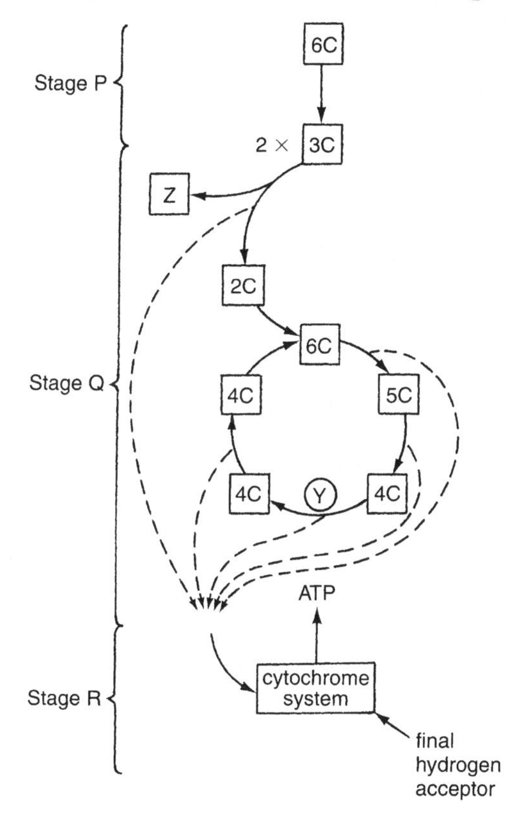

1 Which row in the following table indicates the correct locations at which stages P, Q and R occur in a cell?

	cristae	*cytoplasm*	*central cavity of mitochondrion*
A	R	P	Q
B	Q	P	R
C	P	R	Q
D	R	Q	P

2 Substance Z is

A water.

B lactic acid.

C carbon dioxide.

D adenosine diphosphate.

3 The enzyme which controls reaction Y is a

A decarboxylase. **B** phosphorylase.

C peroxidase. **D** dehydrogenase.

4 The final hydrogen acceptor in the cytochrome system is

A water. **B** oxygen. **C** coenzyme. **D** ADP.

5 During aerobic respiration of one molecule of glucose, most ATP is synthesised during

A glycolysis.
B Krebs' cycle.
C hydrogen transfer along the cytochrome system.
D breakdown of pyruvic acid to a 2-carbon compound.

6 How many molecules of ATP are synthesised as a result of the complete oxidation of one molecule of glucose?

A 2 **B** 4 **C** 36 **D** 38

7 Which of the following are BOTH required by a living cell for glycolysis to occur?

A glucose and oxygen
B ATP and glucose
C oxygen and ATP
D pyruvic acid and oxygen

8 The enzymes required for the Krebs' cycle in a plant cell are located in the

A cytoplasmic fluid surrounding each mitochondrion.
B cristae of each mitochondrion.
C outer membrane of each mitochondrion.
D central matrix of each mitochondrion.

Items 9 and 10 refer to the following diagram of a mitochondrion.

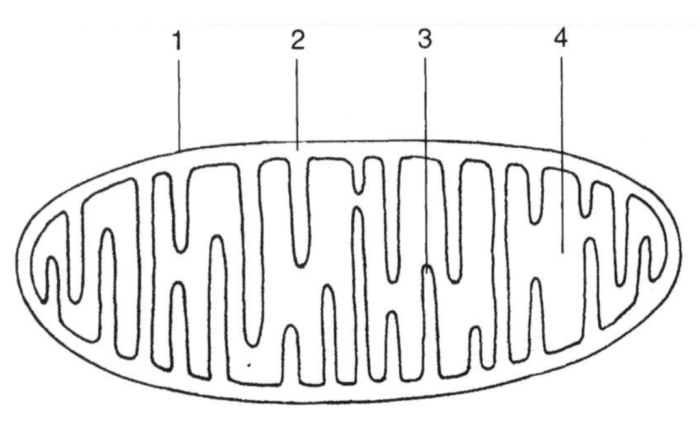

9 Stalked particles would be found at position

A 1 **B** 2 **C** 3 **D** 4

10 Which of the following is LEAST likely to be found in region 4?
A glucose **B** ATP **C** citric acid **D** ADP

11 Which of the following would possess FEWEST mitochondria per unit volume of cell?
A motile sperm cell
B cheek epithelial cell
C nerve cell
D liver cell

12 To respire anaerobically, a yeast cell needs
A alcohol.
B glucose.
C lactic acid.
D cytochrome.

Items **13** and **14** refer to the following possible answers.
A ethanol + CO_2 + ATP
B ethanol + ADP
C lactic acid + CO_2 + ADP
D lactic acid + ATP

13 Which answer correctly identifies the end products resulting from anaerobic respiration in a water-logged root cell?

14 Which answer correctly identifies the end products resulting from anaerobic respiration in mammalian muscle tissue?

15 In an investigation into aerobic respiration in yeast cells, four test tubes were set up as indicated in the following table (where ✓ = present and ✗ = absent).

substance	*test tube*			
	1	2	3	4
live yeast	✓	✗	✓	✗
dead yeast	✗	✗	✗	✓
glucose solution	✗	✓	✓	✓
resazurin dye	✓	✓	✓	✓

The disappearance of resazurin's blue colour will
A occur in tube 3 more quickly than in tube 1.
B occur in tube 2 more quickly than in tube 1.
C fail to occur in both tubes 1 and 2.
D fail to occur in both tubes 3 and 4.

Items **16, 17** and **18** refer to the following experiment which was set up to measure a grasshopper's rate of respiration.

After 30 minutes the coloured liquid in the experiment was returned to its original level by depressing the syringe plunger from point X to point Y.

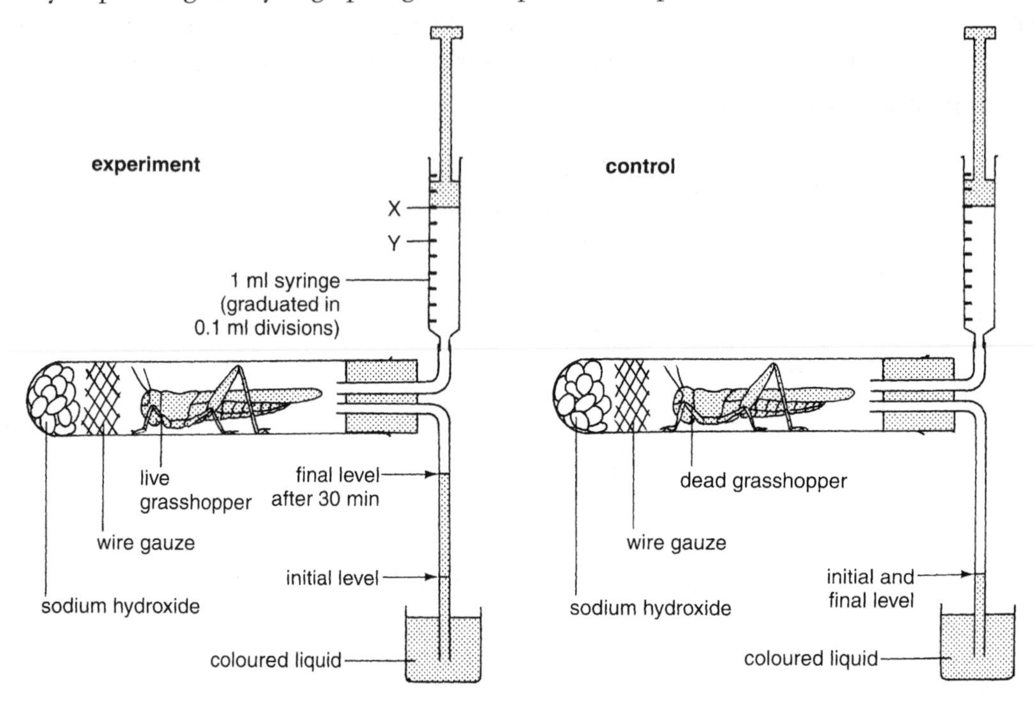

16 The rise in level of coloured liquid indicates that the
- **A** grasshopper is giving out carbon dioxide.
- **B** sodium hydroxide is releasing carbon dioxide.
- **C** grasshopper is taking in oxygen.
- **D** sodium hydroxide is absorbing oxygen.

17 From this experiment it can be concluded that the grasshopper's rate of
- **A** oxygen consumption is 2.0 ml/hour.
- **B** oxygen consumption is 0.4 ml/hour.
- **C** carbon dioxide output is 0.2 ml/hour.
- **D** carbon dioxide output is 4.0 ml/hour.

18 Which of the following procedures would improve the reliability of the result?
- **A** replacing the dead grasshopper in the control tube with glass beads
- **B** using a variety of insects with a period of acclimatisation allowed between readings
- **C** pooling class results where each group uses an adult locust
- **D** repeating the experiment with the same grasshopper and calculating an average

19 If an animal of mass 7 g consumes 3.5 cm^3 of oxygen in 25 minutes, then its respiratory rate in cm^3 oxygen used per gram of body tissue per minute is
A 0.02. **B** 0.08. **C** 0.20. **D** 0.80.

20 Plant cells respire 24 hours per day but only photosynthesise when light is present. The accompanying graph refers to a sample of *Elodea* (Canadian pond weed) over a period of 24 hours.

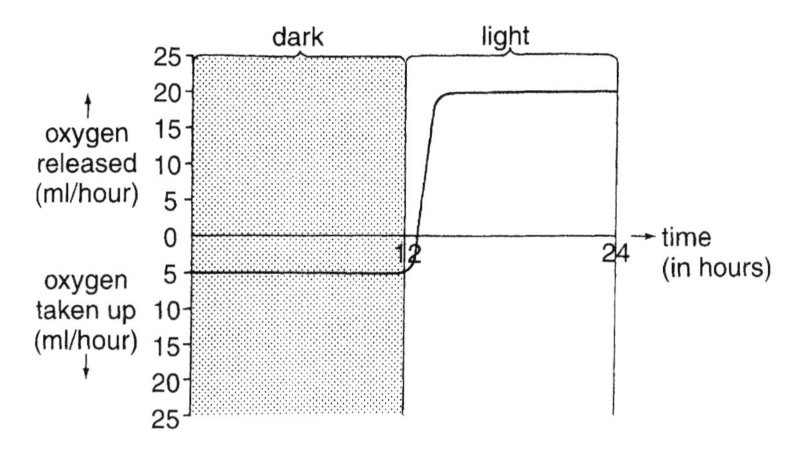

The most accurate estimate of the total volume of oxygen used by the plant for respiration during this 24-hour period is

A 5 ml. **B** 20 ml.
C 120 ml. **D** 160 ml.

5

Role of photosynthetic pigments

Match the terms in list X with their descriptions in list Y.

list X	list Y
1 absorption	**a** green pigments which absorb light in the red and blue regions of the spectrum
2 absorption spectrum	**b** colourless background material in a chloroplast which is rich in enzymes
3 action spectrum	**c** chemical element essential for the formation of chlorophyll molecules
4 carotenoids	**d** technique used to separate the components of a mixture that differ in their degree of solubility in a solvent
5 chlorophylls	**e** process by which light passes through a leaf
6 chloroplast	**f** stack of flattened sacs containing photosynthetic pigments in a chloroplast
7 chromatography	**g** graph showing the rate of photosynthesis by a green plant at different wavelengths of light
8 granum	**h** graph showing the quantity of light absorbed by a pigment at different wavelengths
9 magnesium	**i** process by which light bounces off a leaf surface
10 reflection	**j** discus-shaped organelle found in the cytoplasm of green plant cells
11 stroma	**k** process by which light is taken in and retained by a leaf
12 transmission	**l** accessory yellow pigments (carotene and xanthophyll) which absorb light and pass energy to chlorophyll

Choose the ONE correct answer to each of the following multiple choice questions.

1 Of the total amount of solar energy falling on a leaf, the percentage used for photosynthesis is approximately

A 5%. **B** 25%. **C** 50%. **D** 75%.

2 In the diagram of sunlight striking a green leaf, which arrow represents light being transmitted by the leaf?

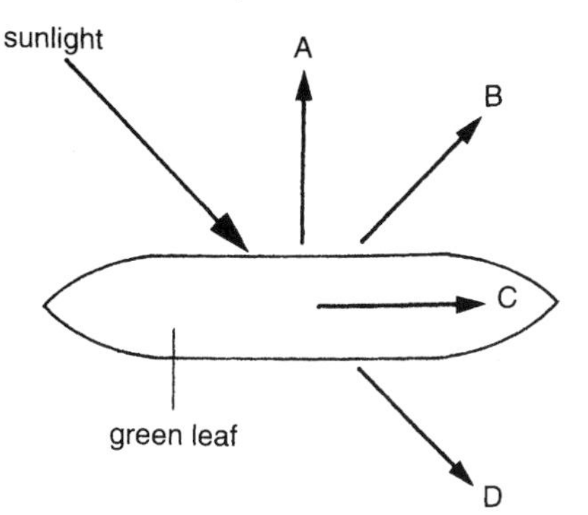

Items 3, 4, 5 and 6 refer to the experiment shown in the following diagram.

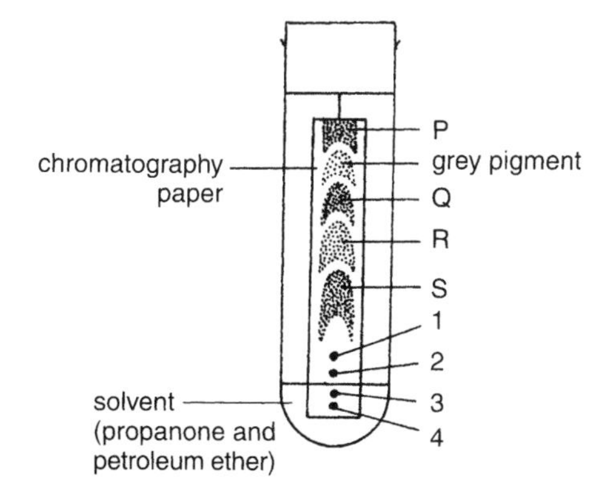

3 Two alternative positions at which the chlorophyll extract could have been spotted to give this separation are

A 1 and 2. **B** 1 and 3.
C 2 and 3. **D** 2 and 4.

4 Xanthophyll is at position

A P. **B** Q. **C** R. **D** S.

5 Chlorophyll is at position

A P. **B** Q. **C** R. **D** S.

6 Carotene is at position
A P. **B** Q. **C** R. **D** S.

7 Which of the accompanying diagrams best represents the absorption spectrum that results when a chlorophyll extract is placed in a beam of white light?

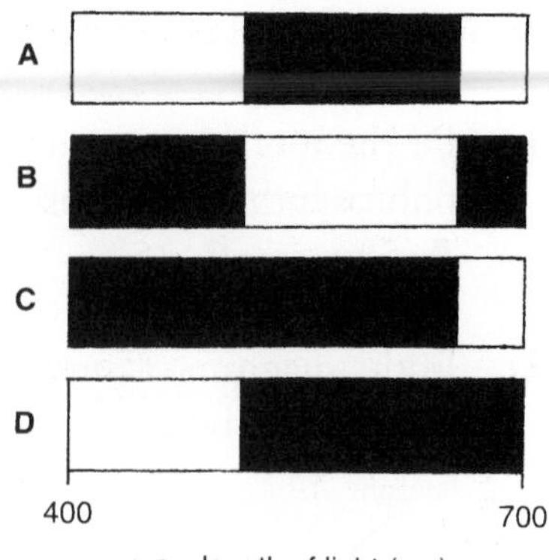

8 Which of the following graphs best represents the absorption spectrum of BOTH green chlorophyll (Ch) and the yellow pigments (Y)?

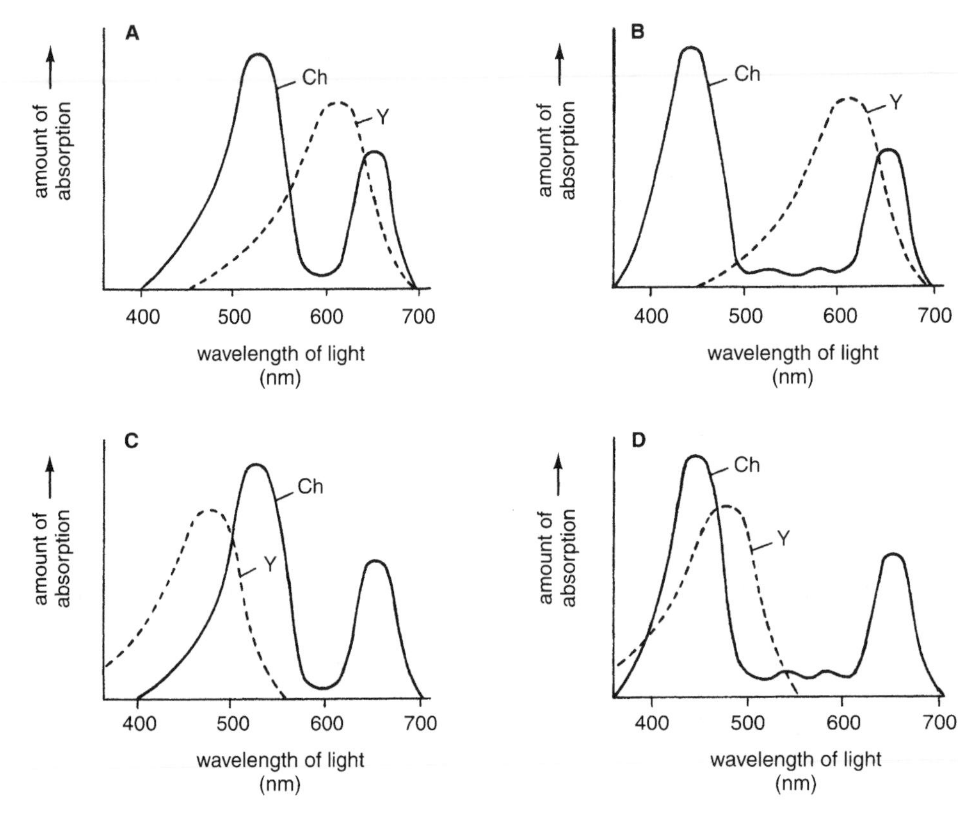

9 The two regions of the spectrum from which chlorophyll absorbs most light are
A blue and yellow. **B** green and blue.
C red and green. **D** red and blue.

10 The diagram opposite shows the result of an experiment in which a strand of alga was placed in water containing bacteria and illuminated by a microspectrum of white light.

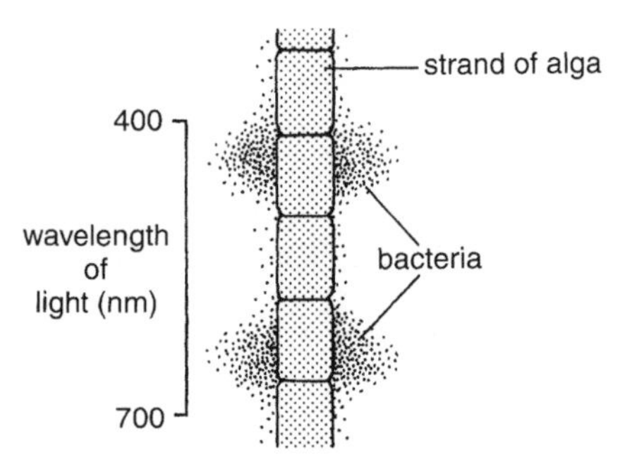

Which of the following correctly explains the distribution of the bacteria?

A The bacteria feed on the alga.
B The alga receives carbon dioxide at these positions.
C The bacteria receive oxygen at these positions.
D The alga feeds on the bacteria.

11 Which of the following metal ions forms part of the structure of a chlorophyll molecule?

A calcium **B** copper **C** iron **D** magnesium

Items **12, 13** and **14** refer to the following diagram which shows the structure of a chloroplast as revealed by an electron microscope.

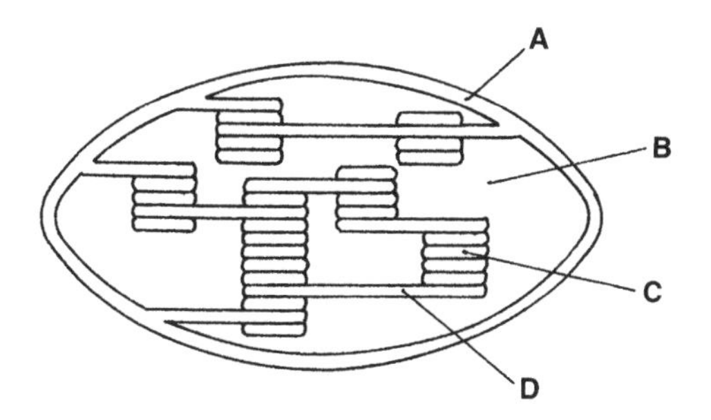

12 Which label line is pointing to the stroma?

13 Which labelled structure is the site of chlorophyll formation?

14 Which letter indicates the site of the carbon fixation stage of photosynthesis?

15 The diagram below shows a plant subjected in turn to four different sets of environmental conditions.

Under which set of conditions would most photosynthesis occur?

D
10°C
daylight

6

Chemistry of photosynthesis

Match the terms in list X (which refer to the first stage in the chemistry of photosynthesis) with their descriptions in list Y.

list X	list Y
1 ADP + Pi	**a** product of the photolysis of water which is required for aerobic respiration
2 chlorophyll	**b** raw material which becomes split into oxygen and hydrogen during photosynthesis
3 granum	**c** compound which accepts hydrogen during the photolysis of water
4 hydrogen	**d** production of ATP using some of the energy trapped during the light-dependent reaction
5 light-dependent reaction	**e** region of a chloroplast containing molecules of chlorophyll
6 NADP	**f** components of a high energy compound formed by photophosphorylation
7 oxygen	**g** breakdown of water during the light-dependent stage of photosynthesis
8 photolysis	**h** green pigment which traps light energy
9 photophosphorylation	**i** product of the photolysis of water which becomes attached to NADP
10 water	**j** first stage in photosynthesis during which light energy is converted to chemical energy

Match the terms in list X (which refer to the second stage in the chemistry of photosynthesis) with their descriptions in list Y.

list X		list Y	
1	ATP	**a**	non-green region of a chloroplast containing enzymes
2	carbon dioxide	**b**	discus-shaped organelle responsible for photosynthesis
3	carbon fixation	**c**	reduced hydrogen acceptor needed for the fixation of carbon in carbohydrate
4	chloroplast	**d**	5-carbon compound which acts as the carbon dioxide acceptor
5	glucose	**e**	3-carbon sugar formed from glycerate phosphate
6	$NADPH_2$	**f**	first stable compound formed in the Calvin cycle after carbon dioxide combines with its acceptor molecule
7	glycerate phosphate	**g**	6-carbon sugar molecule formed from two molecules of triose phosphate
8	ribulose bisphosphate	**h**	raw material which supplies carbon atoms to be fixed into carbohydrate
9	stroma	**i**	high energy compound which provides the energy needed to drive the Calvin cycle
10	triose phosphate	**j**	second stage in photosynthesis which is also known as the Calvin cycle

Choose the ONE correct answer to each of the following multiple choice questions.

1 The oxygen released by a green plant as a result of photolysis comes from
A air. **B** glucose. **C** water. **D** carbon dioxide.

2 The light-dependent stage of photosynthesis results in the formation of two compounds needed for the carbon fixation stage. These are
A adenosine triphosphate and reduced hydrogen acceptor.
B reduced hydrogen acceptor and glycerate phosphate.
C glycerate phosphate and ribulose bisphosphate.
D ribulose bisphosphate and adenosine triphosphate.

3 Photophosphorylation is the name given to the process by which
A chemical energy is converted into light energy in grana.
B ADP and inorganic phosphate are formed by the breakdown of ATP.
C light energy is absorbed by photosynthetic pigments in grana.
D ATP is synthesised during the light-dependent stage of photosynthesis.

4 The first stable compounds resulting from the carbon fixation stage of photosynthesis are formed in the order

A glycerate phosphate → triose phosphate → hexose.
B hexose → glycerate phosphate → triose phosphate.
C glycerate phosphate → hexose → triose phosphate.
D hexose → triose phosphate → glycerate phosphate.

5 The graph below refers to an experiment involving a species of alga. The relative concentrations of GP and RuBP present in the cells were monitored when the plants were in light and then in darkness.

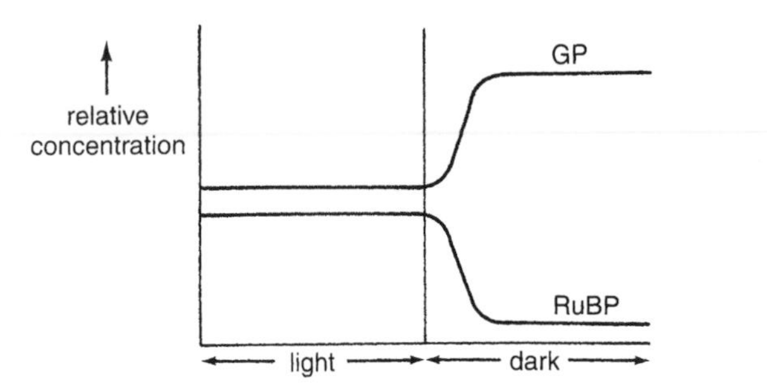

Which of the following conclusions CANNOT be drawn from these results?

A In darkness the relative concentration of GP increases.
B During the experiment RuBP may be converted into GP.
C The relative concentration of RuBP decreases on removal of CO_2.
D In light a steady state exists between RuBP and GP.

Items **6**, **7**, **8**, **9**, **10** and **11** refer to the following diagram of the cyclic series of reactions that occurs during the carbon fixation stage of photosynthesis.

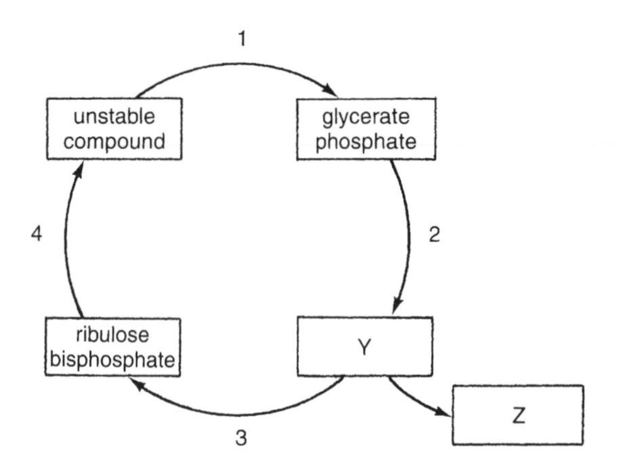

6 This metabolic pathway is also known as the
A tricarboxylic acid cycle.
B citric acid cycle.
C Krebs' cycle.
D Calvin cycle.

7 Carbon dioxide is taken into the cycle at stage
A 1. **B** 2. **C** 3. **D** 4.

8 Hydrogen from reduced hydrogen acceptor is used at stage
A 1. **B** 2. **C** 3. **D** 4.

9 Energy from ATP is used to drive stages
A 1 and 2. **B** 2 and 3. **C** 2 and 4. **D** 3 and 4.

10 The substance formed at position Y is
A 3-carbon sugar. **B** pyruvic acid.
C glucose-1-phosphate. **D** citric acid.

11 If one molecule of substance Y is released per cycle, how many times must the cycle turn for one molecule of sucrose ($C_{12}H_{22}O_{11}$) to be built up at position Z?
A 2 times **B** 4 times **C** 8 times **D** 12 times

12 Which of the following changes in concentration of chemicals would occur if an illuminated green plant cell's source of carbon dioxide were removed?

	ribulose bisphosphate	*glycerate phosphate*
A	increase	increase
B	decrease	decrease
C	increase	decrease
D	decrease	increase

13 The accompanying graph presents the results from a photosynthesis experiment.

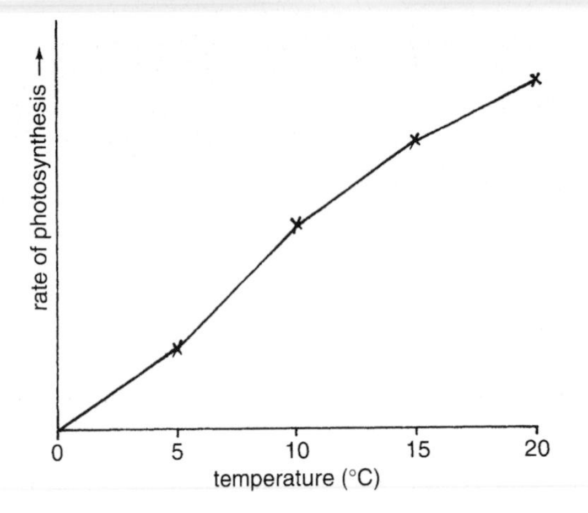

Which of the following pairs of environmental factors must have been kept constant to ensure the validity of the experiment?

A light intensity and temperature
B temperature and carbon dioxide concentration
C water content and oxygen concentration
D light intensity and carbon dioxide concentration

Items **14** and **15** refer to the graph below which shows the effect of carbon dioxide concentration on the rate of photosynthesis at three different intensities of light.

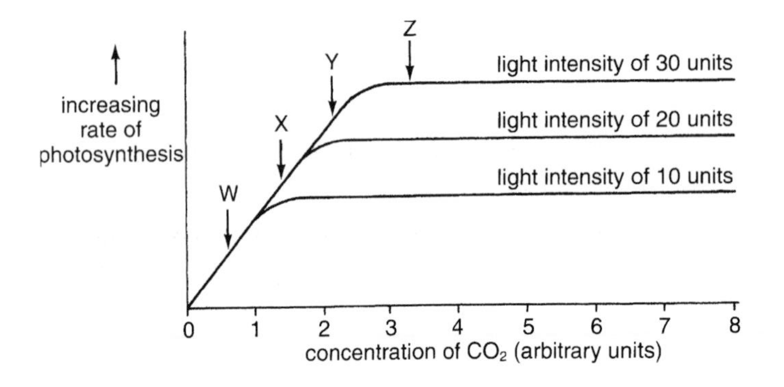

14 At which of the following concentrations was carbon dioxide always the limiting factor?

A 0–1 units **B** 1–2 units **C** 2–3 units **D** 3–4 units

15 Light intensity was the limiting factor at

A point W only. **B** point Z only.
C points X and Y only. **D** points W, X, Y and Z.

7

DNA and its replication

Match the terms in list X with their descriptions in list Y.

list X		**list Y**	
1	adenine	**a**	two-stranded molecule of DNA wound up into a spiral
2	cytosine	**b**	process by which a molecule of DNA reproduces itself
3	deoxyribose	**c**	base present in DNA that is complementary to thymine
4	deoxyribonucleic acid	**d**	basic unit of which nucleic acids are composed
5	DNA polymerase	**e**	sugar present in DNA
6	double helix	**f**	base present in DNA that is complementary to adenine
7	guanine	**g**	base present in DNA that is complementary to guanine
8	nucleotide	**h**	nucleic acid present in chromosomes
9	replication	**i**	base present in DNA that is complementary to cytosine
10	thymine	**j**	enzyme required to promote DNA replication

Choose the ONE correct answer to each of the following multiple choice questions.

1 DNA is NOT present in a

A nucleus. **B** gene.
C membrane. **D** chromosome.

2 The sugar present in DNA is

A ribose. **B** dextrose.
C ribulose. **D** deoxyribose.

3 The structure of one nucleotide is shown below.

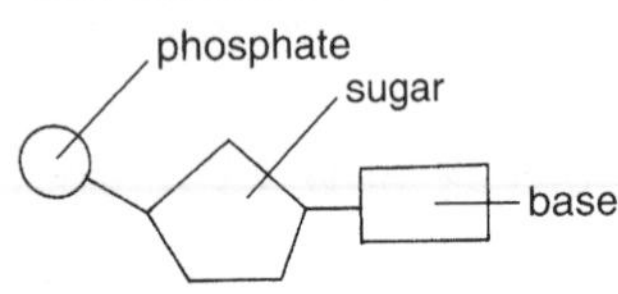

Which of the following diagrams shows two nucleotides correctly joined together?

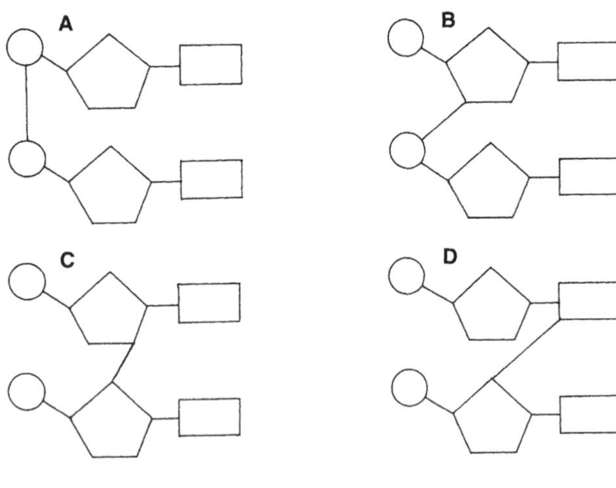

4 Which of the following is a base pair normally present in DNA?

A adenine and cytosine **B** guanine and adenine
C thymine and guanine **D** thymine and adenine

Items 5 and 6 refer to the following table and list of possible answers.

cell types analysed	*average mass of DNA/cell ($\times 10^{-12}$ g)*
X	0.00
Y	3.35
kidney	6.70
lung	6.70

A sperm cell **B** liver cell
C smooth muscle cell **D** mature red blood cell

5 What is the correct identity of cell type X?

6 What is the correct identity of cell type Y?

7 The average mass of DNA present in an ovum of the species referred to in the above table would be

A 3.35×10^{-6} **B** 6.70×10^{-6}
C 3.35×10^{-12} **D** 6.70×10^{-12}

8 If a DNA molecule contains 10 000 base molecules of which 18% are thymine, then the number of cytosine molecules present is

A 1800 **B** 3200 **C** 6400 **D** 8200

9 A shorthand method of representing part of a single strand of DNA is shown opposite.

Which of the following shows the correct positions of the phosphate (P), sugar (S) and base (B) molecules?

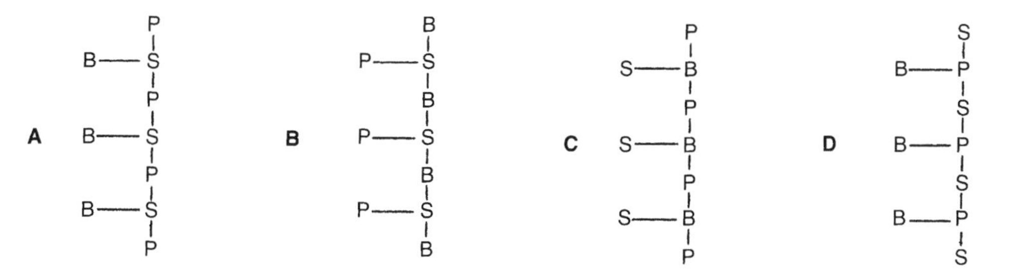

10 If a DNA molecule contains 4000 base molecules and 1200 of these are adenine, then the percentage number of guanine bases present in the molecule is

A 12 **B** 20 **C** 28 **D** 30

11 DNA molecules isolated from a rat cell and human cell are found to differ in the order of their

A bases only. **B** sugars only.
C phosphates only. **D** bases, sugars and phosphates.

12 The following set of results shows an analysis of the DNA bases contained in the cells of a cow's thymus gland.

base composition			
X	*guanine*	Y	Z
28.2%	21.5%	21.2%	27.8%

Which of the following is a possible correct identification of the bases?

	X	Y	Z
A	cytosine	adenine	thymine
B	thymine	adenine	cytosine
C	adenine	cytosine	thymine
D	cytosine	thymine	adenine

13 During DNA replication the following events occur:

1 winding brings about formation of two double helices
2 bases on free nucleotides bond with bases on the DNA strand
3 hydrogen bonds break allowing DNA strands to unzip
4 bonds form between adjacent nucleotide molecules

The correct order in which these events occur is
A 2, 3, 1, 4. **B** 3, 4, 2, 1. **C** 2, 3, 4, 1. **D** 3, 2, 4, 1.

14 The following diagram shows a molecule of DNA prior to replication.

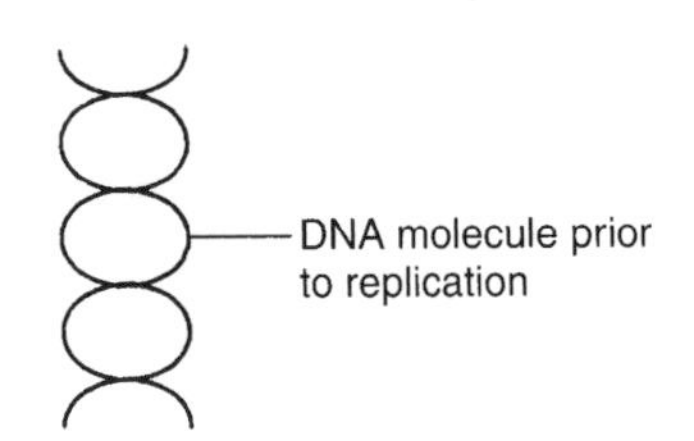

If ———— represents an original DNA strand and ------------ represents a new DNA strand, which of the following daughter DNA molecules will result from replication of the DNA molecule shown above?

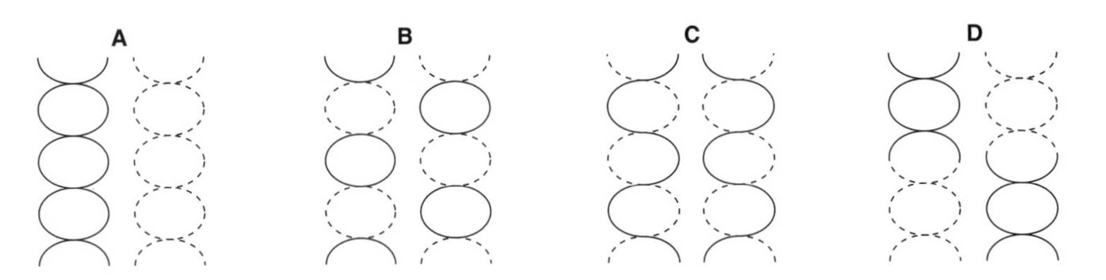

15 The diagram below shows the stages that occur in an actively dividing mammalian cell.

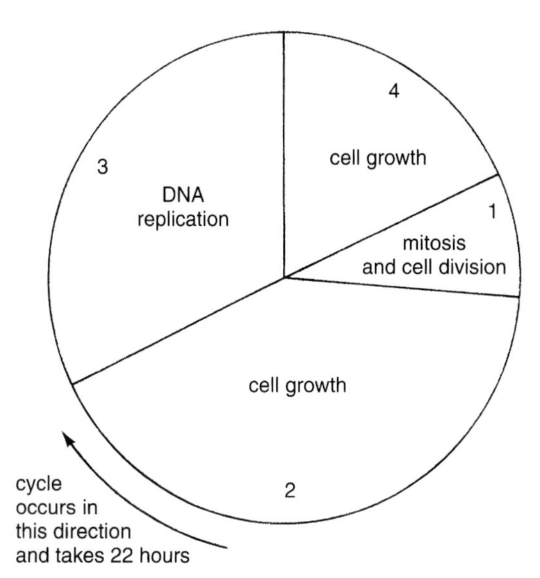

If the drug aminopterin (which inhibits thymine formation) is added to a culture of actively dividing cells, at which stage in the cell cycle will most cells be present 16 hours after the addition of the drug?

A 1 or 2 **B** 2 or 3 **C** 3 or 4 **D** 4 or 1

8 RNA and protein synthesis

Match the terms in list X with their descriptions in list Y.

list X		**list Y**	
1	amino acid	**a**	sub-cellular structure which is the site of protein synthesis
2	anticodon	**b**	process by which a complementary molecule of mRNA is made from a region of a DNA template
3	codon	**c**	type of nucleic acid which conveys information from DNA to a ribosome
4	genetic code	**d**	long chain of amino acids formed at a ribosome during translation of RNA
5	Golgi apparatus	**e**	the conversion of the genetic code into a sequence of amino acids in a polypeptide
6	mRNA	**f**	molecular language made up of 64 codewords called codons
7	polypeptide	**g**	unit of genetic code consisting of three mRNA bases
8	ribose	**h**	type of nucleic acid which acts as an amino acid carrier
9	ribosome	**i**	triplet of bases on a tRNA molecule which is complementary to a mRNA codon
10	rough endoplasmic reticulum	**j**	one of 20 different types of sub-unit which make up protein molecules
11	transcription	**k**	sugar present in RNA
12	translation	**l**	base present in RNA that is complementary to adenine
13	tRNA	**m**	network of flattened sacs and tubes with ribosomes attached to membranes
14	uracil	**n**	system of flattened sacs from which secretory vesicles arise in a cell

Choose the ONE correct answer to each of the following multiple choice questions.

1 Which of the following is correct?

	present in DNA	*present in RNA*
A	uracil	thymine
B	deoxyribose	ribose
C	single strand	double strand
D	four different nucleotides	five different nucleotides

2 One of the nucleotides present in mRNA has the composition

A adenine – ribose – phosphate.
B uracil – deoxyribose – phosphate.
C thymine – ribose – phosphate.
D guanine – deoxyribose – phosphate.

3 Strand X in the diagram shows a small part of a nucleic acid molecule.

strand X: A, C, A, G, T

Which pair of the following strands are complementary to strand X?

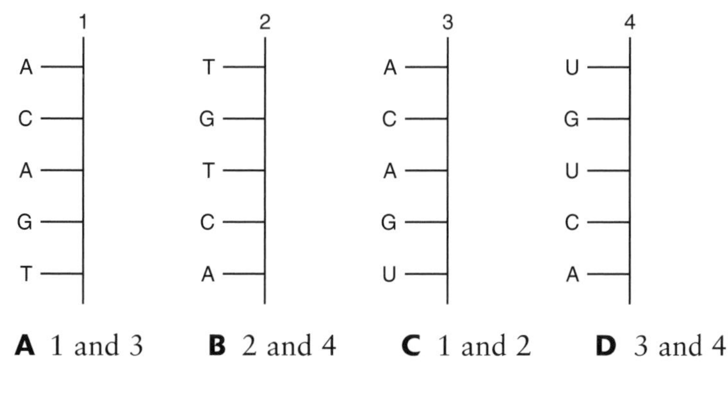

A 1 and 3 **B** 2 and 4 **C** 1 and 2 **D** 3 and 4

4 A mRNA template is

A translated from protein. **B** transcribed into protein.
C translated in DNA. **D** transcribed from DNA.

5 A free transfer RNA molecule can combine with

A one specific amino acid only. **B** any available amino acid.
C three different amino acids. **D** a chain of amino acids.

6 The number of bases present in one codon is

A 1 **B** 2 **C** 3 **D** 4

Items 7, 8, 9 and 10 refer to the possible answers numbered in the following list.

1 DNA molecules **2** tRNA molecules **3** mRNA molecules
4 amino acids **5** ribosomes

7 On which of these are anticodons present?
A 2 **B** 3 **C** 4 **D** 5

8 On which of these are base triplets found which are complementary to anticodons?
A 1 **B** 2 **C** 3 **D** 4

9 Which of these must be present in the largest number for successful synthesis of a large protein molecule to occur?
A 1 **B** 3 **C** 4 **D** 5

10 Which of these controls the order in which amino acids are added to a growing protein chain?
A 2 **B** 3 **C** 4 **D** 5

11 If each amino acid weighs 100 mass units, what is the weight (in mass units) of the protein molecule synthesised from a mRNA molecule which is 600 bases long?
A 2000 **B** 6000 **C** 20 000 **D** 60 000

Items 12, 13 and 14 refer to the following diagram which shows the synthesis of part of a protein molecule.

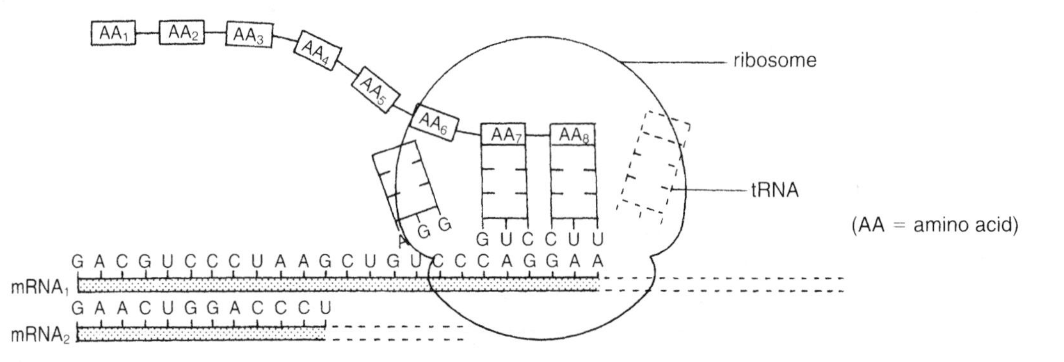

12 Which of the following is the first part of the protein molecule that would be translated from $mRNA_2$?

start of protein
↓

A $AA_4 - AA_2 - AA_7 - AA_6$ _ _ _ _ _ _
B $AA_6 - AA_7 - AA_2 - AA_4$ _ _ _ _ _ _
C $AA_3 - AA_1 - AA_5 - AA_8$ _ _ _ _ _ _
D $AA_8 - AA_5 - AA_1 - AA_3$ _ _ _ _ _ _

13 The DNA strand from which $mRNA_2$ was synthesised is

A GAACTGGACCCT **B** CTTGACCTGGGA
C GAACUGGACCCU **D** CUUGACCUGGGA

14 The following diagram shows a small part of a different protein that was also synthesised on this ribosome.

AA_7 – AA_2 – AA_4 – AA_6 –

What sequence of bases in DNA coded for this sequence of amino acids?

A CAGGUCAAGUCC **B** CAGCTCAAGTCC
C GUCCAGUUCAGG **D** GTCCAGTTCAGG

15 Ribosomes are found to occur

A freely in the cytoplasm and attached to the endoplasmic reticulum.
B attached to the endoplasmic reticulum and to the nuclear membrane.
C freely in the cytoplasm and attached to the nuclear membrane.
D attached to the Golgi apparatus and to the endoplasmic reticulum.

16 Which of the following is composed of a system of flattened sacs and tubules?

A nuclear membrane **B** mitochondrion
C endoplasmic reticulum **D** ribosome

17 In which of the following would the Golgi apparatus be most highly developed?

A salivary gland cells **B** red blood corpuscles
C skeletal muscle fibres **D** kidney tubule cells

18 The relative number of ribosomes would be greatest in

A sieve tubes in a ripe fruit.
B xylem vessels in a woody stem.
C epidermal cells in a mature root.
D leaf cells in a growing bud.

19 Which of the following statements about the Golgi apparatus is INCORRECT?

A It is a group of flattened fluid-filled sacs.
B It is formed from vesicles pinched off from the endoplasmic reticulum.
C It contains newly synthesised protein.
D It is attached to the nuclear membrane.

20 Prior to mucus (a type of protein) leaving a goblet cell and playing its role in the trachea, the following events occur.

1 fusion of vesicle with plasma membrane
2 addition of carbohydrate component to protein
3 secretion of processed glycoprotein by cell
4 separation of vesicle from Golgi apparatus

The sequence in which these events occur is

A 4, 2, 3, 1. **B** 2, 4, 1, 3. **C** 4, 2, 1, 3. **D** 2, 4, 3, 1.

9

Functional variety of proteins

Match the terms in list X with their descriptions in list Y.

list X		**list Y**	
1	amino acid	**a**	molecule consisting of polypeptide chains folded into a ball and associated with a non-protein chemical
2	conjugated protein	**b**	one of 20 different types of organic compound which are the basic building blocks of protein
3	fibrous protein	**c**	weak chemical link holding a polypeptide chain in a coil by linking adjacent amino acids in the helix
4	globular protein	**d**	molecule consisting of long parallel polypeptide chains arranged like a rope
5	hydrogen bond	**e**	strong chemical link joining adjacent amino acids in a polypeptide chain
6	nitrogen	**f**	molecule consisting of polypeptide chains arranged like a ball of string
7	peptide bond	**g**	chain-like molecule composed of several amino acids
8	polypeptide	**h**	chemical element present in protein but absent from carbohydrates

Choose the ONE correct answer to each of the following multiple choice questions.

1 Which of the following chemical elements is present in proteins but not in fats?
A carbon **B** oxygen **C** nitrogen **D** hydrogen

2 Which of the following chemical elements is often a constituent of protein?
A calcium **B** sodium **C** potassium **D** sulphur

3 The number of different types of amino acid commonly found to make up proteins is approximately
A 20 **B** 30 **C** 200 **D** 1000

Items 4 and 5 refer to the following possible answers.
A fibrous protein **B** hydrolysed protein
C conjugated protein **D** globular protein

4 Which of these consists of polypeptide chains arranged in long parallel strands?

5 Which of these is illustrated by the following diagram?

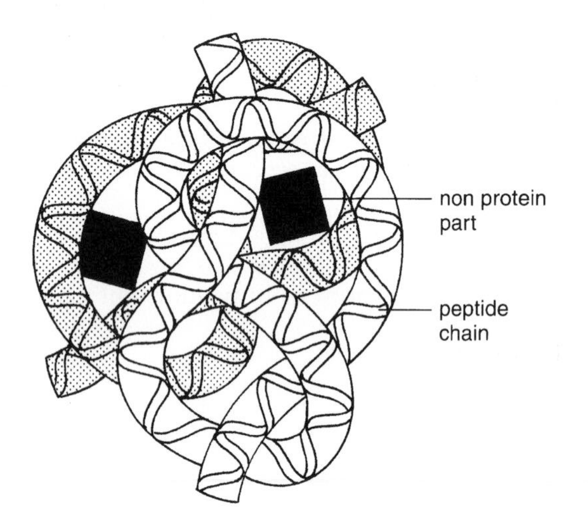

6 Which of the following is an example of a conjugated protein?
A insulin **B** haemoglobin
C pepsin **D** glucagon

7 Three proteins found in the human body are
1 thyroxine
2 cytochrome oxidase
3 immunoglobulin (antibody)

Which of the following correctly identifies their functions?

	function		
	regulates growth and metabolism	*defends the body against microbes*	*speeds up rate of a biochemical process*
A	1	2	3
B	2	3	1
C	1	3	2
D	3	1	2

8 The following table gives the mass per 100 g of protein of five different amino acids found in four proteins.

	protein	*mass of amino acid (g/100 g protein)*				
		glycine	*alanine*	*leucine*	*valine*	*phenylalanine*
A	insulin	4	5	13	8	8
B	haemoglobin	6	7	15	9	8
C	keratin	7	4	11	5	4
D	albumin	3	7	9	7	8

Which protein is represented by the following pie chart?

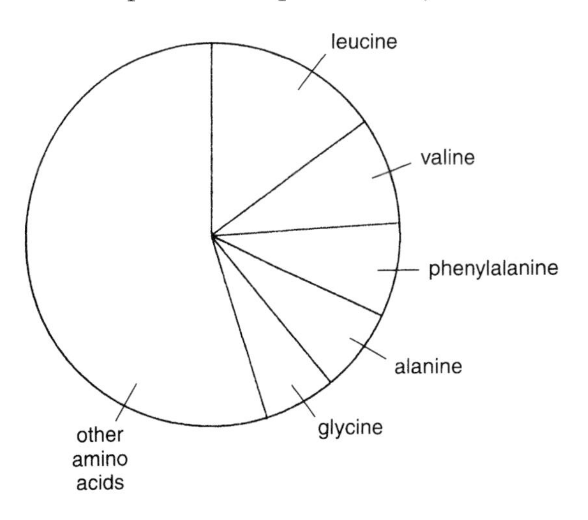

9 The diagram opposite shows the sequence of amino acids present in one molecule of insulin.

In this protein the ratio of leucine: glycine: tyrosine: histidine is

A 6:4:3:2 **B** 6:4:4:1
C 3:2:1:1 **D** 3:2:2:1

Glycine		Phenylalanine
Isoleucine		Valine
Valine		Asparagine
Glutamic acid		Glutamic acid
Glutamic acid		Histidine
Cysteine		Leucine
S Cysteine	—S–S—	Cysteine
Alanine		Glycine
Serine		Serine
S Valine		Histidine
Cysteine		Leucine
Serine		Valine
Leucine		Glutamic acid
Tyrosine		Alanine
Glutamic acid		Leucine
Leucine		Tyrosine
Glutamic acid		Leucine
Asparagine		Valine
Tyrosine	S–	Cysteine
Cysteine	—S	Glycine
Asparagine		Glutamic acid
		Arginine
		Glycine
		Phenylalanine
		Phenylalanine
		Tyrosine
		Threonine
		Proline
		Lysine
		Alanine

10 In the following diagram of enzyme action, which structure represents the enzyme?

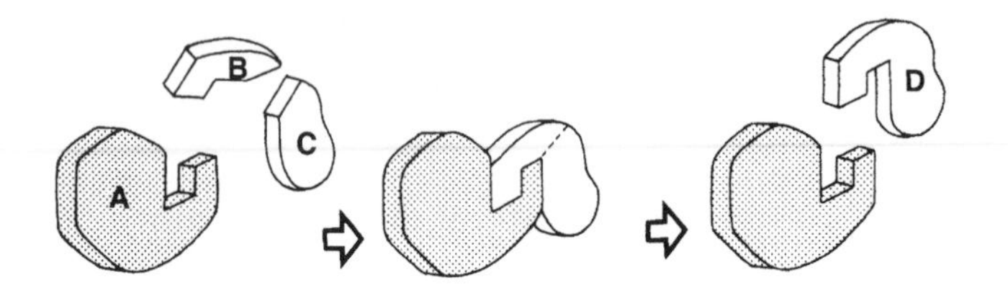

10 Viruses

Match the terms in list X with their descriptions in list Y.

list X

1 AIDS

2 bacteriophage

3 capsid

4 helper T cell

5 HIV

6 host

7 lysis

8 retrovirus

9 reverse transcriptase

10 virus

list Y

a general name for a micro-organism that exhibits living and non-living characteristics

b bursting of host cell releasing copies of virus

c virus which destroys helper T cells

d general name for virus containing RNA and reverse transcriptase

e protective coat surrounding a virus

f enzyme which transcribes viral RNA into DNA

g type of virus that attacks bacteria

h type of white blood cell that plays host to HIV

i organism whose cells are attacked by a virus

j acquired immune deficiency syndrome

Choose the ONE correct answer to each of the following multiple choice questions.

Questions **1** and **2** refer to the accompanying diagram of a virus.

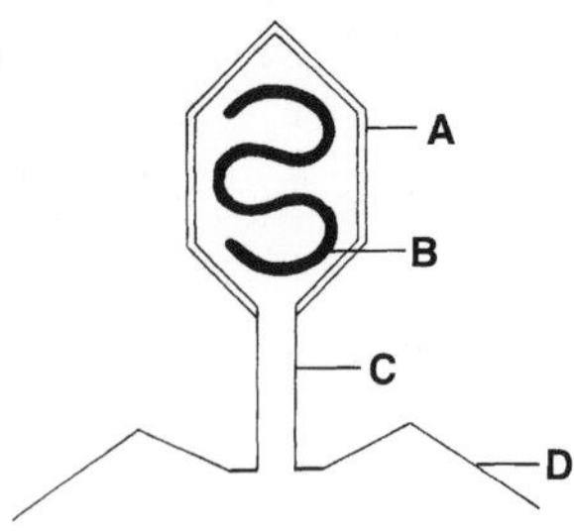

1 Which labelled structure is nucleic acid?

2 Which labelled structure is the protein coat?

3 It is CORRECT to say that viruses
A reproduce both inside and outside living cells.
B exhibit living and non living characteristics.
C are always transmitted from animal to animal by blood contact.
D are always transmitted from plant to plant by aphids.

4 The diameter of a certain virus is 20 nanometres (nm). If 1 micrometre (μm) = 1000 nanometres, then the diameter of this virus expressed in metres (m) is
A 2×10^{-8} **B** 2×10^{8} **C** 2×10^{-9} **D** 2×10^{9}

5 In the following comparison, which pair of statements is NOT correct?

	virus	*unicellular alga*
A	contains one type of nucleic acid	contains two types of nucleic acid
B	depends on host cell for synthesis of nucleic acid and protein coats	is able to synthesise all of its own nucleic acid and protein molecules
C	cannot form the first link in a food chain	can form the first link in a food chain
D	always arises directly from another virus	never arises directly from a pre-existing cell

Items **6** and **7** refer to the following diagram which shows some of the stages that occur during viral invasion of a bacterium.

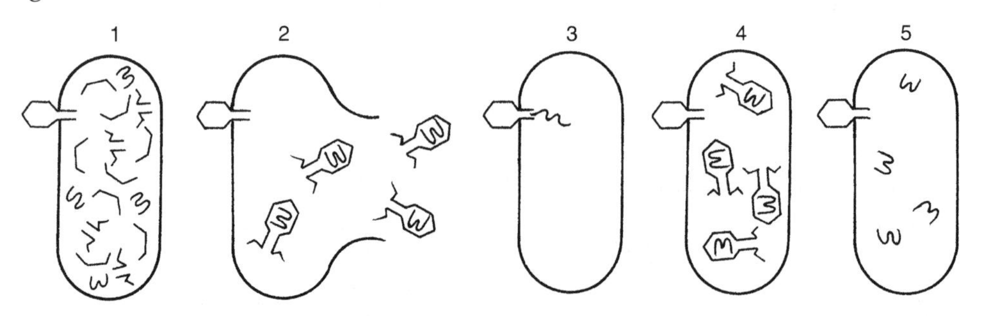

6 The sequence in which these events would occur is

A 3, 1, 5, 4, 2. **B** 2, 4, 1, 5, 3.

C 3, 5, 1, 4, 2. **D** 2, 5, 1, 3, 4.

7 Which numbered stage illustrates viral nucleic acid molecules about to become surrounded by newly formed protein coats?

A 1 **B** 2 **C** 4 **D** 5

8 Which numbered stage shows the host cell undergoing lysis?

A 1 **B** 2 **C** 3 **D** 4

Items **9**, **10**, **11**, **12** and **13** refer to the branched key below which was constructed using the information in the following table.

virus	*nucleic acid*	*protein coat*
coliphage fd	single-stranded DNA	naked and helical
pox virus	double-stranded DNA	enveloped and helical
herpes virus	double-stranded DNA	enveloped and polyhedral
adenovirus	double-stranded DNA	naked and polyhedral
tobacco mosaic	single-stranded RNA	naked and helical
picornavirus	single-stranded RNA	naked and polyhedral
myxovirus	single-stranded RNA	enveloped and helical
reovirus	double-stranded RNA	naked and polyhedral

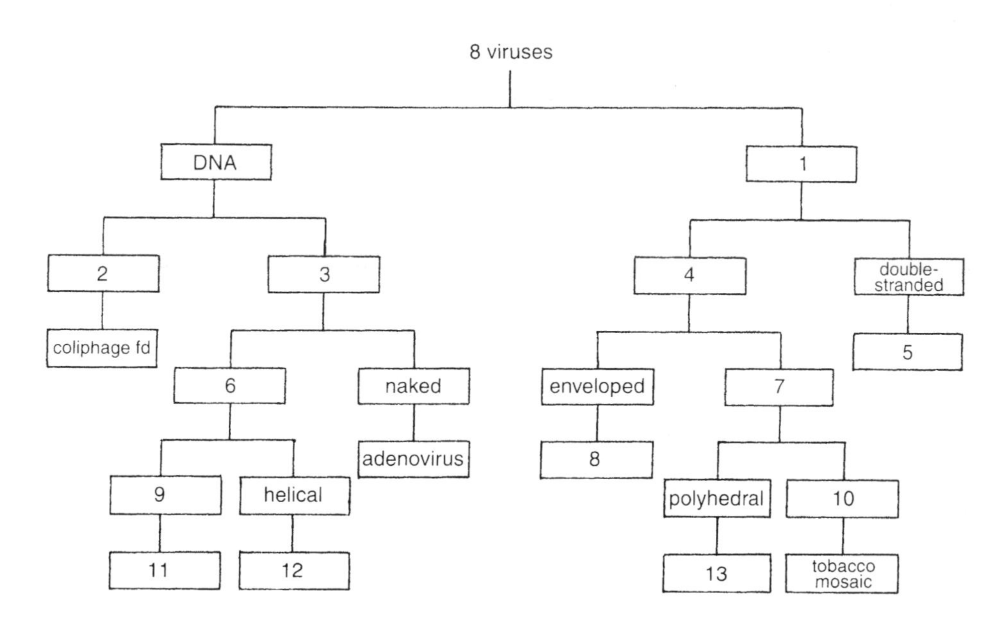

9 The characteristic 'double-stranded' should be at position
A 1 **B** 2 **C** 3 **D** 4

10 The characteristic 'helical' should be at position
A 6 **B** 7 **C** 9 **D** 10

11 Pox virus would be correctly classified at position
A 5 **B** 8 **C** 11 **D** 12

12 The virus at position 13 should be
A herpes. **B** myxovirus.
C picornavirus. **D** reovirus.

13 The identity of the virus shown in the accompanying diagram is
A tobacco mosaic. **B** picornavirus.
C myxovirus. **D** reovirus.

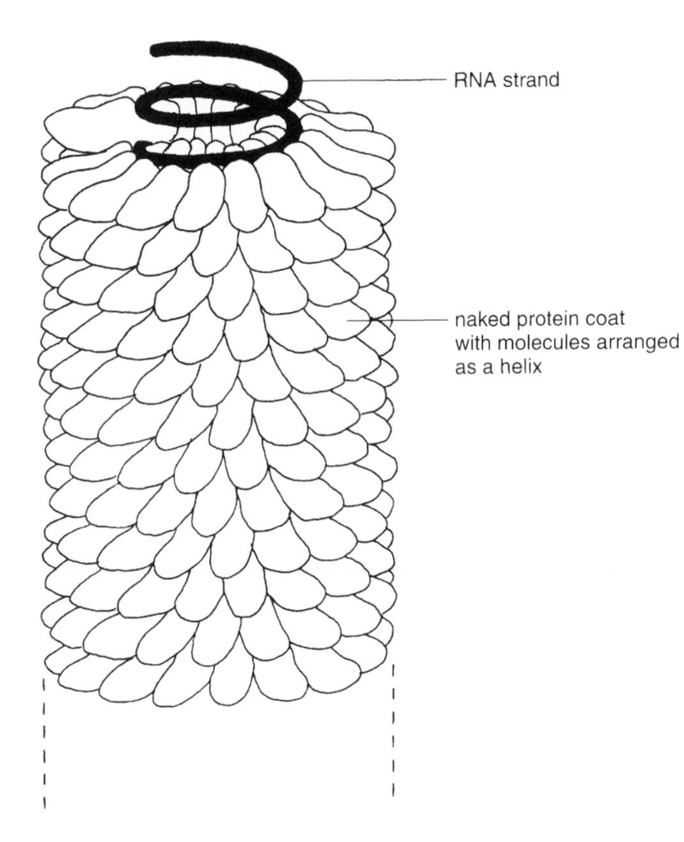

14 A retrovirus contains

A RNA and ATP.

B DNA and RNA.

C RNA and reverse transcriptase.

D DNA and reverse transcriptase.

15 Which of the following statements is TRUE of HIV (human immunodeficiency virus)?

A HIV becomes attached to its host cell by reverse transcriptase.

B Helper T cells in human blood are attacked by HIV.

C HIV is caused by a retrovirus called AIDS.

D Copies of HIV are released from a host cell by lysis.

11
Cellular defence mechanisms

Match the terms in list X with their descriptions in list Y.

list X		list Y	
1	active immunity	**a**	type of white blood cell which makes antibodies
2	antibody	**b**	sub-cellular structure containing enzymes employed during phagocytosis
3	antigen	**c**	protection gained by receiving ready-made antibodies from another person or animal
4	cyanide	**d**	type of white blood cell which engulfs and digests bacteria
5	lymphocyte	**e**	sticky acidic substance made by plants to isolate an area of injury
6	lysosome	**f**	protection gained by producing antibodies in response to natural or artificial exposure to a pathogen
7	nicotine	**g**	poison made by plant cells which blocks the cytochrome system of an invading pathogen
8	passive immunity	**h**	process by which bacteria are engulfed and destroyed by certain white blood cells
9	phagocyte	**i**	non-sticky acidic chemical made by plant cells which inhibits activity of a pathogen's enzymes
10	phagocytosis	**j**	complex molecule recognised by lymphocytes as foreign to the body
11	resin	**k**	protein molecule specific to one antigen
12	tannin	**l**	natural insecticide produced by tobacco plant cells

Choose the ONE correct answer to each of the following multiple choice questions.

1 Ideal conditions for the growth of pathogens in the human body are provided by a supply of

A food, carbon dioxide and warmth.
B moisture, food and carbon dioxide.
C warmth, moisture and food.
D carbon dioxide, moisture and warmth.

2 The following diagram shows four of the stages which occur during the process of phagocytosis.

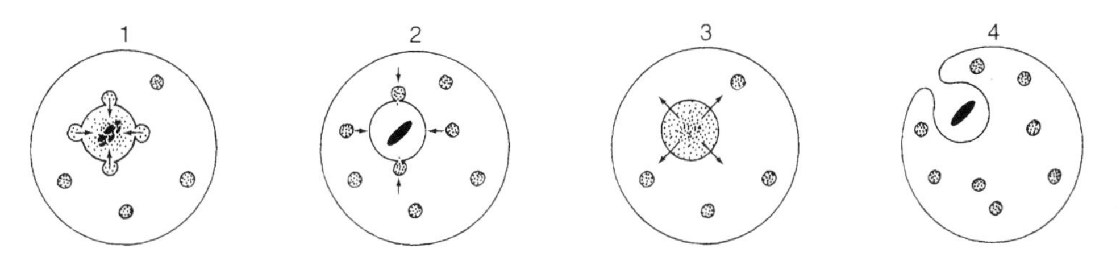

Which set in the table below correctly matches each of these numbered stages with its description?

description of stage	*set*			
	A	B	C	D
products of digestion pass into cytoplasm of phagocyte	3	4	2	1
some lysosomes move towards and fuse with vacuole	2	1	3	4
phagocyte forms vacuole around bacterium	4	3	1	2
digestive enzymes break bacterium down	1	2	4	3

3 Which of the diagrams below shows the correct structure of an antibody molecule? (R = receptor site, N = non-receptor site)

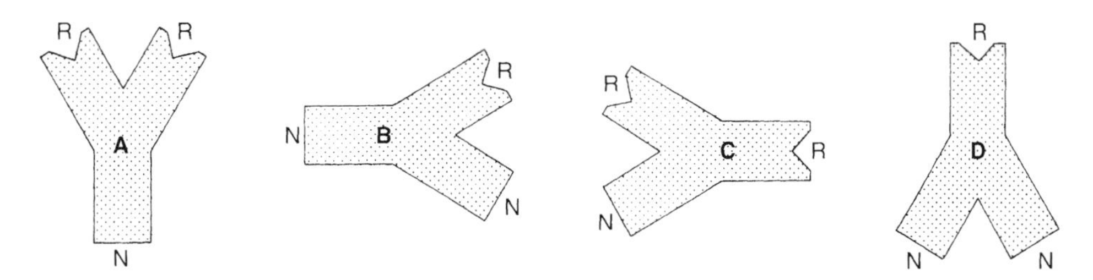

4 Which of the following statements is CORRECT?
A An antibody stimulates the body to produce antigens.
B An antibody is always composed of phospholipids.
C An antigen stimulates white blood cells to make antibodies.
D An antigen is always composed of viral protein.

5 When a virus with antigens on its surface invades an organism and multiplies, the following events occur.
1 many copies of antibodies synthesised by white blood cells
2 virus particles become attached by antigens to white blood cells
3 virus antigens combine with antibodies
4 white blood cells multiply rapidly

The order in which these occur is
A 3, 2, 1, 4. **B** 2, 4, 1, 3.
C 4, 1, 3, 2. **D** 2, 1, 4, 3.

Items 6 and 7 refer to the accompanying diagram of types of specific immunity.

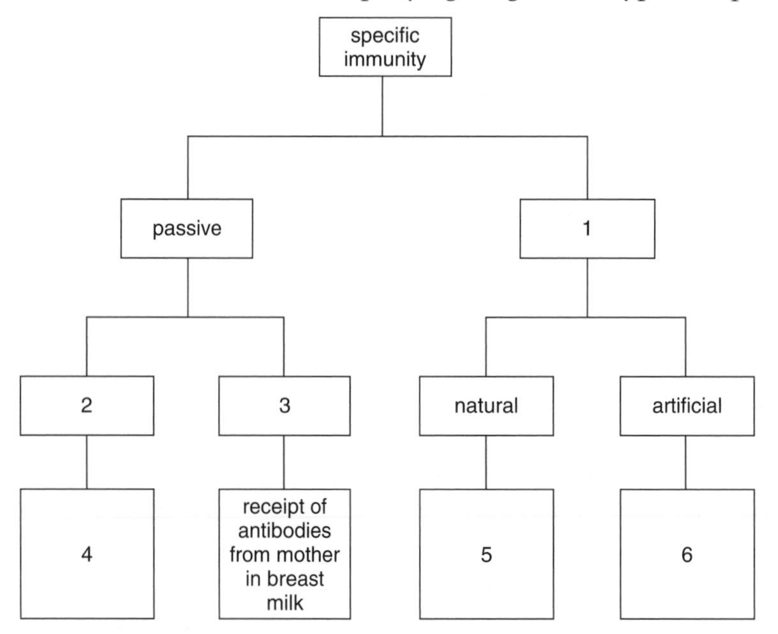

6 The example 'production of memory cells following infection by chicken pox virus' should have been inserted in box
A 1. **B** 3. **C** 5. **D** 6.

7 The example 'receipt of tetanus antibodies made by a horse' should have been inserted in box
A 2. **B** 4. **C** 5. **D** 6.

8 Which of the following graphs best represents the primary and secondary responses on exposure to a pathogen?

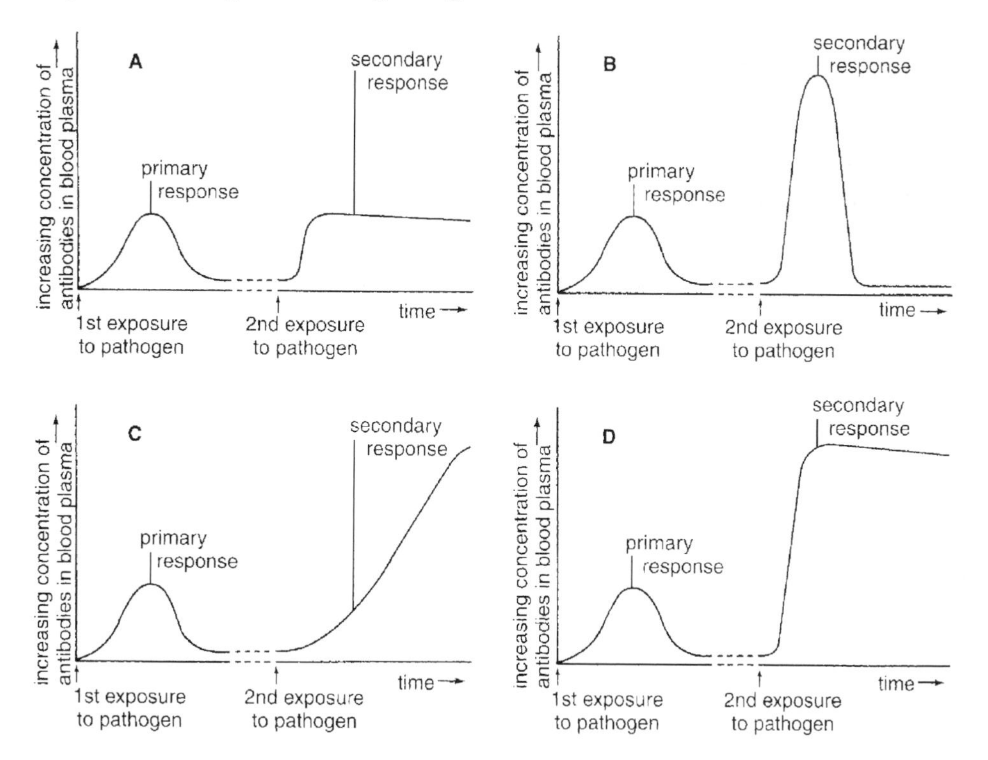

9 Compared with the secondary response, antibody formation during the primary response is normally found to

A reach a higher level.
B be maintained for a longer time.
C occur at a much more rapid rate.
D be unable to prevent the disease.

10 The following statements refer to the process of tissue transplantation in the human body. Which statement about the recipient's immune system is FALSE?

A It regards the transplanted tissue as a collection of foreign antigens.
B It would try to destroy the transplanted tissue from the donor.
C It is deliberately inhibited by suppressors to prevent tissue rejection.
D Its suppression makes the recipient resistant to many serious infections.

11 The following diagram shows the crosses which resulted in the production of three mice (1, 2 and 3).

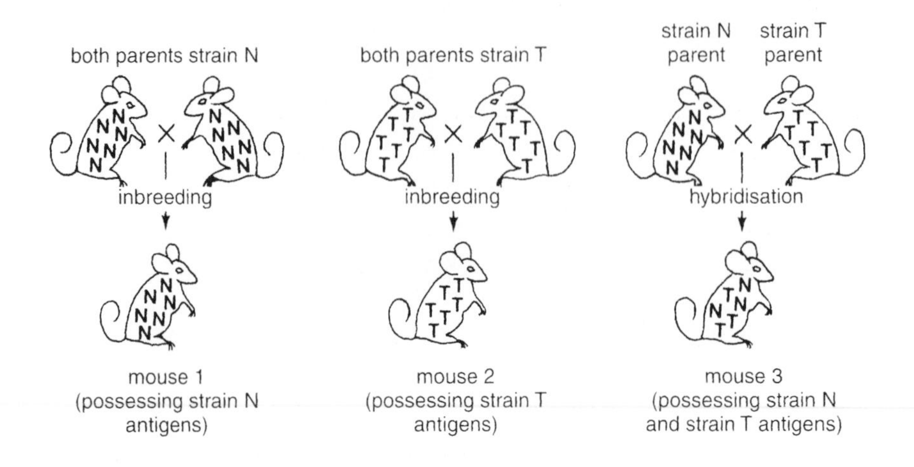

Which of the following skin grafts would be LEAST likely to be rejected?

	donor mouse	*recipient mouse*
A	3	1
B	2	1
C	1	3
D	3	2

Items **12**, **13** and **14** refer to the following chemicals made by certain plants as mechanisms of cellular defence.

A tannin **B** cyanide **C** nicotine **D** resin

12 Which type of chemical is used to isolate an injured area of a plant under attack by a pathogen?

13 Which type of chemical acts as a poison by blocking the invader's cytochrome system?

14 Which type of chemical occurs in the vacuoles and cell walls of many plant organs and disturbs an invader's metabolism by acting as an enzyme inhibitor?

15 Which of the following equations correctly represents the process of cyanogenesis?

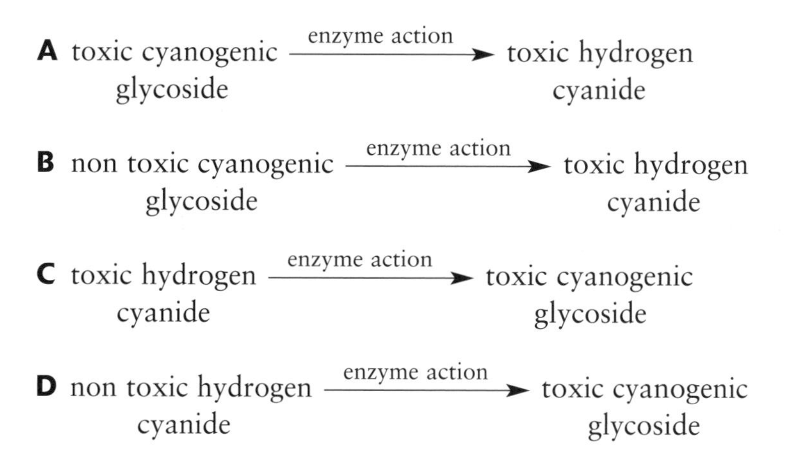

A toxic cyanogenic glycoside $\xrightarrow{\text{enzyme action}}$ toxic hydrogen cyanide

B non toxic cyanogenic glycoside $\xrightarrow{\text{enzyme action}}$ toxic hydrogen cyanide

C toxic hydrogen cyanide $\xrightarrow{\text{enzyme action}}$ toxic cyanogenic glycoside

D non toxic hydrogen cyanide $\xrightarrow{\text{enzyme action}}$ toxic cyanogenic glycoside

Items **16, 17, 18** and **19** refer to the graph below. It charts the quantities of two types of chemical present in bracken leaf fronds which begin to push up out of the soil in spring.

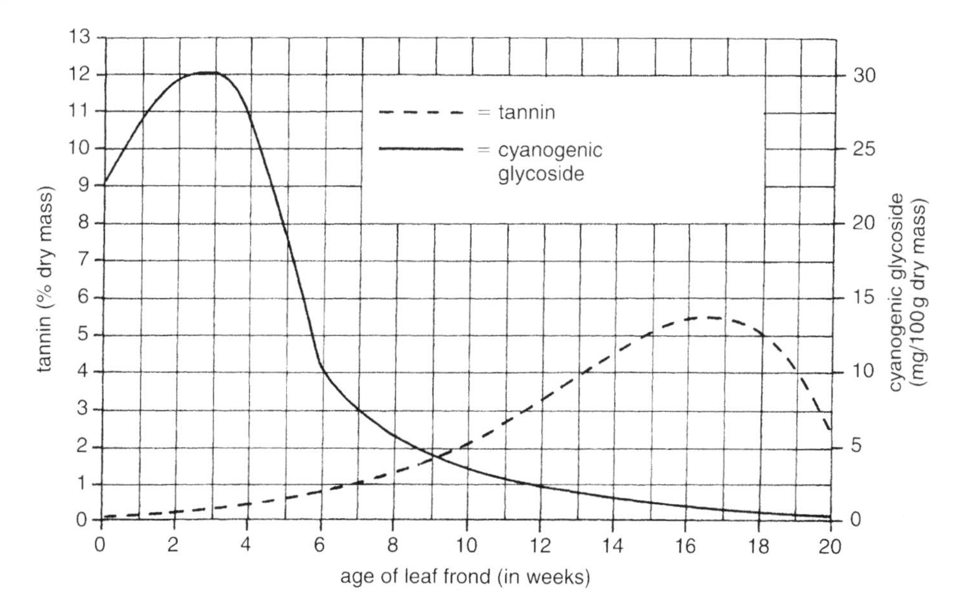

16 Which of the following conclusions can be correctly drawn from the graph?

A Younger leaves possessed a high concentration of cyanogenic glycoside and a low concentration of tannin.

B Older leaves possessed a high concentration of cyanogenic glycoside and a low concentration of tannin.

C Younger leaves possessed a high concentration of both cyanogenic glycoside and tannin.

D Older leaves possessed a high concentration of both cyanogenic glycoside and tannin.

17 The quantities of the two chemicals present in leaves at week 7 were

	tannin (*% dry mass*)	*cyanogenic glycoside* (*mg/100 g dry mass*)
A	1.0	3.0
B	1.0	7.5
C	3.0	2.5
D	7.5	1.0

18 The concentration of cyanogenic glycoside showed the biggest drop in concentration between weeks
A 3 and 4. **B** 4 and 5. **C** 5 and 6. **D** 6 and 7.

19 The percentage decrease in concentration of cyanogenic glycoside that occurred between weeks 3 and 7 was
A 9. **B** 25. **C** 75. **D** 90.

20 The following table summarises a series of experiments involving the sawfly which causes 'bean galls' on the leaves of willow trees.

details of experiment	*result*
fertile female allowed to puncture leaf and lay eggs which hatch	+
fertilised eggs extracted from female and implanted into leaf tissues	−
fertile female allowed to pierce leaf and lay eggs which are then killed using a needle	+
adult males left in contact with gall-free willow leaves	−
fertile female allowed to puncture leaf but prevented from laying eggs	+
hungry larvae removed from galls and placed on gall-free leaf	−
sterilised female allowed to pierce leaf and lay sterile eggs which fail to hatch	+

(+ = gall produced, − = no gall produced)

Which of the following hypotheses is supported by this series of experiments?
In this particular relationship between gall-causer and gall-maker, gall formation
A is stimulated by a chemical released by the digestive glands of feeding larvae.
B depends on a chemical present on the surface of newly laid eggs.
C only occurs if eggs release a chemical on hatching into larvae.
D is induced by a chemical injected by the female at the time of egg-laying.

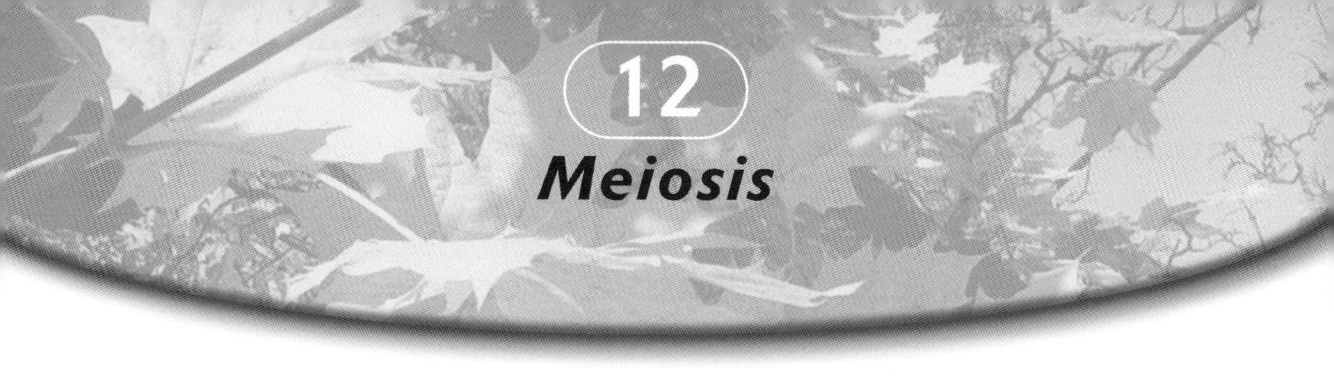

12 Meiosis

Match the terms in list X with their descriptions in list Y.

list X		list Y	
1	centromere	**a**	one of the two longitudinal sub-units of a duplicated chromosome
2	chiasma	**b**	possessing two sets of chromosomes
3	chromatid	**c**	the characteristic number of chromosomes typical of a species
4	chromosome	**d**	cross-shaped arrangement of two chromatids at a point of crossing over
5	chromosome complement	**e**	small region of chromosome which becomes associated with spindle fibres during meiosis
6	crossing over	**f**	unit of heredity occupying a specific site on a chromosome
7	diploid	**g**	a form of nuclear division producing four haploid gametes from a diploid cell
8	gene	**h**	two chromosomes identical in size and matching one another gene for gene (although alleles may differ)
9	haploid	**i**	differences that exist amongst the members of a species that are determined by genes
10	homologous pair	**j**	breaking and rejoining of adjacent chromatids leading to exchange of genetic material
11	independent assortment	**k**	thread-like structure composed of genes and found in the nucleus of a cell
12	inherited variation	**l**	arrangement of homologous pairs of chromosomes that allows their segregation at meiosis
13	meiosis	**m**	possessing one set of chromosomes

Choose the ONE correct answer to each of the following multiple choice questions.

1 The following diagram shows the formation of an animal zygote. This zygote contains

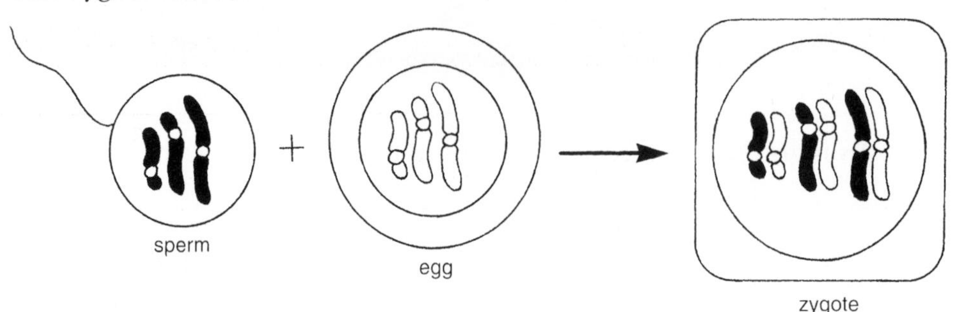

A six chromatids and is haploid.
B six chromatids and is diploid.
C three pairs of chromosomes and is haploid.
D three pairs of chromosomes and is diploid.

2 Meiosis
A involves only one cell division.
B produces identical daughter cells.
C increases variation in a population.
D produces new body cells during repair.

3 Replication of the deoxyribonucleic acid (DNA) necessary for meiosis occurs
A while the chromosomes are arranged at the equator.
B before the chromatids appear.
C between the first and second divisions.
D after the homologous chromosomes become separated.

4 Which of the products shown in the following table results from meiosis in a sperm mother cell (ploidy number = 2n)?

	number of sperm formed	*ploidy number of each sperm*
A	2	n
B	2	2n
C	4	n
D	4	2n

5 The diagram below shows the nuclear contents of a cell. Which of the descriptions in the table refers accurately to this cell?

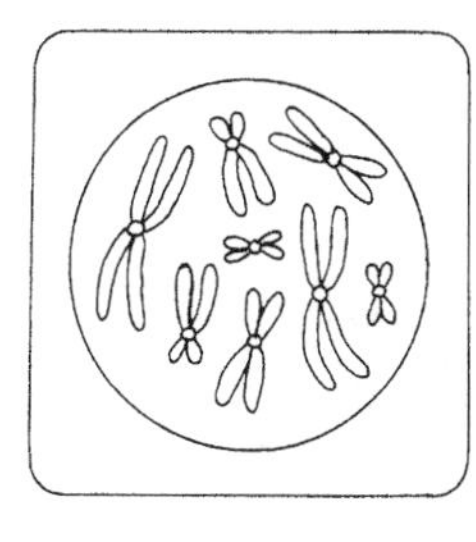

	number of chromosomes	*number of pairs of homologous chromosomes*
A	8	4
B	8	8
C	16	4
D	16	8

6 The following diagram shows a sperm mother cell from a fruit fly. The paternal chromosomes are shaded and the maternal chromosomes are unshaded.

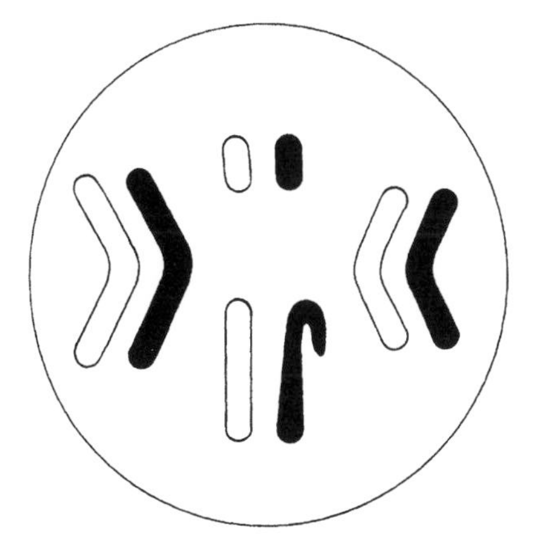

The chance of a sperm receiving all four maternal chromosomes is
A 1 in 2. **B** 1 in 4. **C** 1 in 8. **D** 1 in 16.

7 Which of the following correctly shows a pair of homologous chromosomes at the start of meiosis?

A

B

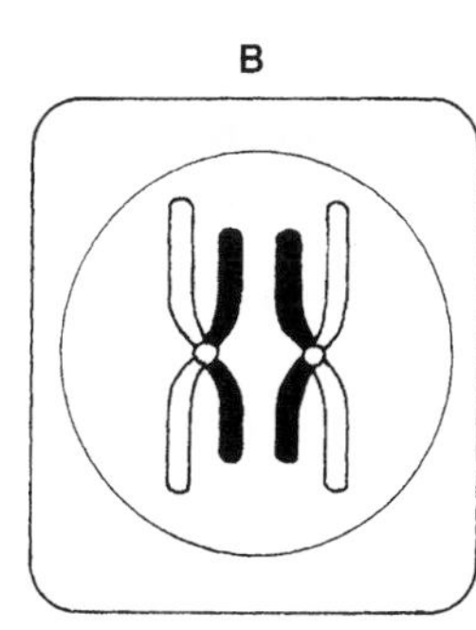

C

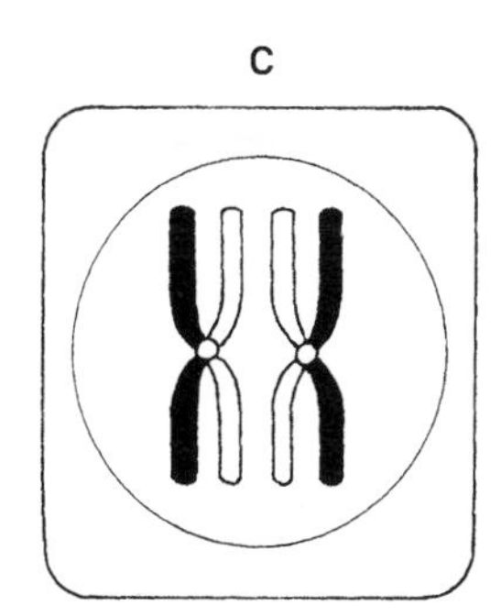

D

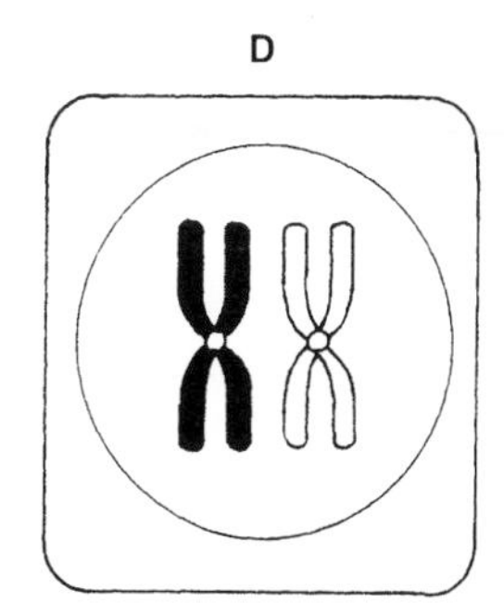

8 The following diagram shows a cell undergoing meiosis.

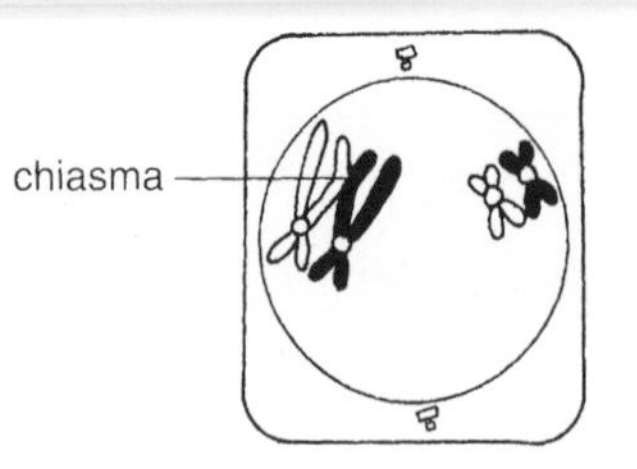

Which of the diagrams below shows the next stage in the process?

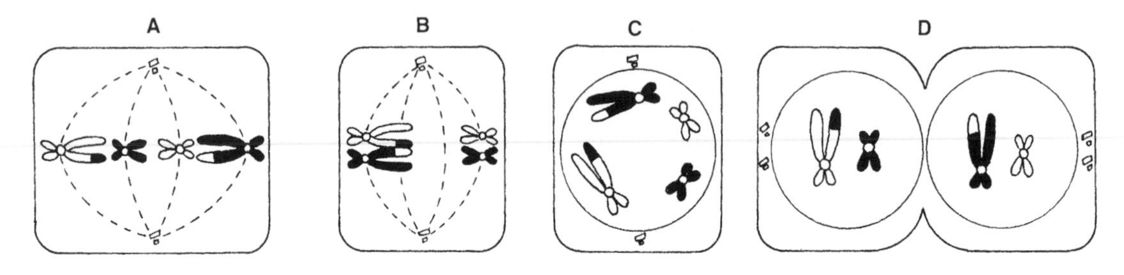

9 In the accompanying diagram of a cell undergoing meiosis, assume that crossing over occurs only at the chiasma indicated.

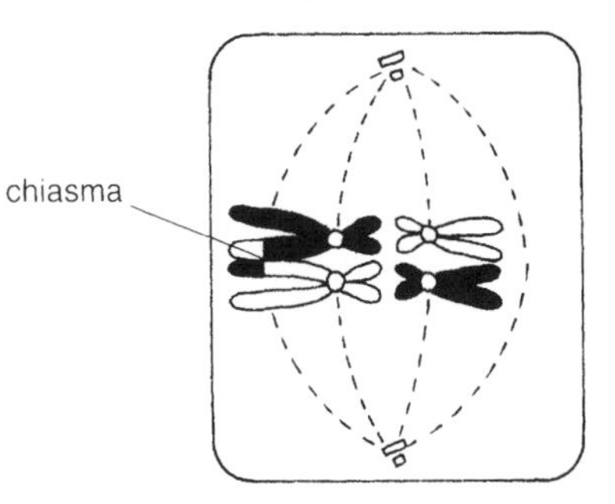

Which of the following gametes will NOT be formed from this cell?

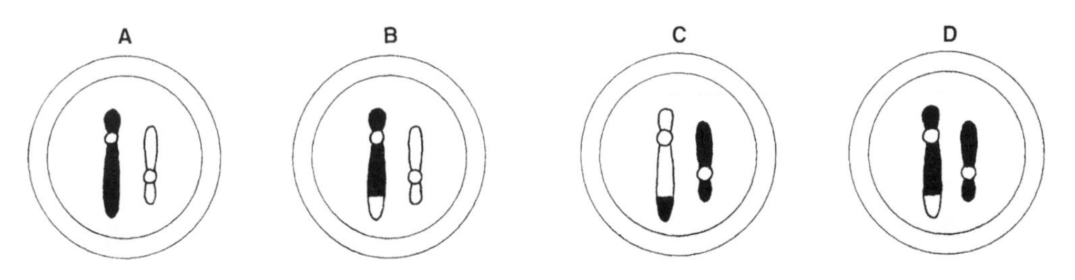

10 In which of the following do BOTH structures contain cells which divide by meiosis?

A stigma and stamen **B** ovary and anther
C stamen and style **D** anther and stigma

11 Which of the following could be used to prepare a microscope slide of cells undergoing meiosis?

A locust testis **B** pea root tip
C human cheek epithelium **D** fruit fly salivary gland

12 The following diagram represents the human life cycle.

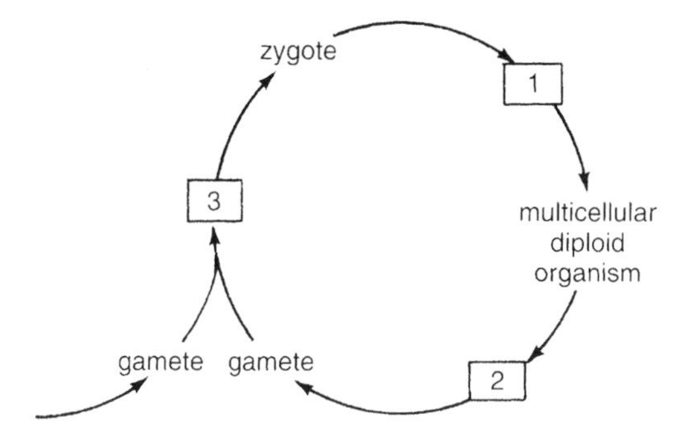

Which of the following combinations of terms correctly matches the numbered boxes?

	1	*2*	*3*
A	mitosis	meiosis	fertilisation
B	meiosis	mitosis	fertilisation
C	fertilisation	meiosis	mitosis
D	mitosis	fertilisation	meiosis

13 Four different steps that occur during meiosis are given in the following list.

1 complete separation of chromatids
2 pairing of homologous chromosomes
3 lining up of paired chromosomes on equator
4 crossing over between chromatids

These steps would occur in the order

A 2, 3, 4, 1. **B** 3, 2, 4, 1. **C** 2, 4, 3, 1. **D** 3, 1, 2, 4.

14 Each egg mother cell in a dandelion plant contains 24 chromosomes. The number of chromosomes present in a root tip cell would be

A 6 **B** 12 **C** 24 **D** 48

15 Which of the following statements is true of mitosis but NOT of meiosis?

A The chromosome number is unaltered by the process.
B The cells produced by the process differ from each other.
C Variation within a population is increased by the process.
D The cells produced by the process are all haploid.

13 Monohybrid cross

Match the terms in list X with their descriptions in list Y.

list X	list Y
1 complete dominance	**a** genotype possessing two different alleles of a gene
2 genotype	**b** form of gene that is masked by the presence of the dominant form
3 heterozygous	**c** breeding experiment involving two true-breeding parents differing in only one way
4 homozygous	**d** condition of heterozygote in which one allele is fully expressed in the phenotype and the other is masked
5 incomplete dominance	**e** the complete set of genes possessed by an organism
6 monohybrid cross	**f** condition of heterozygote in which partial expression of alleles produces a phenotype intermediate between the two homozygotes
7 phenotype	**g** genotype possessing two identical alleles of a gene
8 recessive allele	**h** an organism's appearance resulting from genetic information inherited from parents

Choose the ONE correct answer to each of the following multiple choice questions.

1 In humans, the condition hyperdactyly (the possession of twelve fingers) is determined by a dominant allele (H) and the normal condition by the recessive allele (h).

The following diagram shows a family tree in which some members of the family are hyperdactylous.

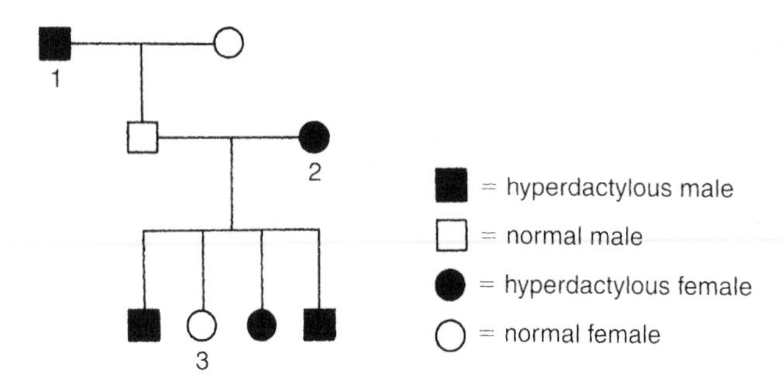

The genotypes of persons 1, 2 and 3 in this family tree are

	1	*2*	*3*
A	HH	Hh	hh
B	HH	HH	hh
C	Hh	HH	Hh
D	Hh	Hh	hh

2 In *Drosophila*, the allele for normal grey body colour (G) is dominant to ebony body (g). The following table summarises the results of several crosses.

cross	*result*
strain 1 × gg	all wild type
strain 2 × gg	1 wild type : 1 ebony
strain 3 × gg	all ebony
strain 4 × Gg	3 wild type : 1 ebony

Which strains BOTH have the genotype Gg?

A 1 and 3
B 1 and 4
C 2 and 3
D 2 and 4

3 In maize plants, two alleles of the gene for seed colour exist. Purple (P) is dominant to yellow (p).

A backcross (testcross) was carried out to determine the genotype of a certain purple plant. Which of the following is correct?

	phenotypic ratio of offspring resulting from backcross	*genotype of purple parent*
A	1 purple : 1 yellow	heterozygous
B	3 purple : 1 yellow	homozygous
C	1 purple : 1 yellow	homozygous
D	all purple	heterozygous

4 In mice, Y is the dominant allele for yellow fur and y is the recessive allele for grey fur. Since Y is **lethal** when homozygous, the result of cross Yy × Yy will be

A 3 yellow : 1 grey. **B** 2 yellow : 1 grey.
C 1 yellow : 1 grey. **D** 1 yellow : 2 grey.

5 A certain type of anaemia exists in two forms, major (severe) and minor (mild). The following table relates the genotypes of both types of sufferer to that of a normal person.

person	*genotype*
non sufferer	NN
minor sufferer	NA
major sufferer	AA

If NA marries NA, the chance of each of their children being mildly affected is

A 1 in 1. **B** 1 in 2. **C** 1 in 3. **D** 1 in 4.

Items 6 and 7 refer to the following family tree.

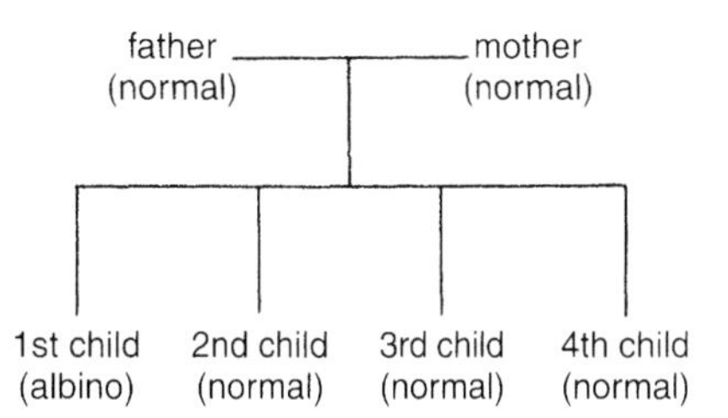

6 If A = normal allele and a = albino allele, the genotypes of these parents are

	father	*mother*
A	Aa	Aa
B	AA	AA
C	AA	Aa
D	Aa	AA

7 The chance of this couple's fifth child being an albino is
A 1 in 2. **B** 1 in 3. **C** 1 in 4. **D** 1 in 5.

8 In snapdragon plants, the alleles for red and ivory flower colour show incomplete dominance.
When a homozygous red-flowered plant is crossed with a homozygous ivory-flowered plant, all the members of the F_1 generation are found to bear pink flowers.
Which of the following would be the outcome of crossing a red-flowered plant with a pink one?
A 1 red : 2 pink : 1 ivory **B** 3 red : 1 ivory
C 1 red : 1 pink **D** all red

9 In mice, black coat colour (allele B) is dominant to brown coat colour (allele b). The offspring of a cross between a black mouse (BB) and a brown mouse were allowed to interbreed. What percentage of the progeny would have black coats?
A 25% **B** 50% **C** 75% **D** 100%

10 In fruit flies, long wing is dominant to vestigial wing. When heterozygous long-winged flies were crossed with vestigial-winged flies, 192 offspring were produced. Of these 101 had long wings and 91 had vestigial wings.

If an exact Mendelian ratio had been obtained, then the number of each phenotype would have been

	long-winged	*vestigial-winged*
A	64	128
B	96	96
C	128	64
D	192	0

14 Dihybrid cross and linkage

Match the terms in list X with their descriptions in list Y.

list X	**list Y**
1 chiasma	**a** name given to the position of a gene on a chromosome
2 crossing over	**b** segregation of different alleles of two genes on different chromosomes during gamete formation
3 dihybrid backcross	**c** point of crossing over between adjacent chromatids during meiosis
4 dihybrid cross	**d** process by which new combinations of parental characteristics occur in a future generation
5 independent assortment	**e** breeding experiment involving heterozygous F_1 organisms and individuals with the double recessive genotype
6 linked genes	**f** form of a species that occurs most commonly in an ecosystem
7 locus	**g** general term for a breeding experiment involving two true-breeding parents differing in two respects
8 recombination	**h** breaking and rejoining of adjacent chromatids leading to exchange of genetic material
9 wild type	**i** units of heredity located on the same chromosome

Choose the ONE correct answer to each of the following multiple choice questions.

Items **1** and **2** refer to the following information and list of possible answers.

In a certain plant, yellow fruit colour (Y) is dominant to green (y) and round shape (R) is dominant to oval (r). The two genes involved are located on different chromosomes.

A 9:3:3:1 ratio of phenotypes only
B 9:3:3:1 ratio of genotypes only
C 1:1:1:1 ratio of phenotypes only
D 1:1:1:1 ratio of phenotypes and genotypes

1 Which of the above will result when plant YyRr is self-pollinated?

2 Which of the above will result when plant YyRr is backcrossed (testcrossed) with the double recessive parent?

Items **3** and **4** refer to the following information.

In *Drosophila*, long wing (L) is dominant to dumpy wing (l) and grey body (G) is dominant to ebony body (g). The two genes involved are not on the same chromosome.

A true-breeding long-winged, ebony-bodied fly is crossed with a true-breeding dumpy-winged, grey-bodied fly.

3 The genotype of the F_1 generation will be

A LlGg **B** LLGg **C** LLGG **D** LlGG

4 As a result of interbreeding amongst the members of the F_1 generation, dumpy-winged, grey-bodied flies will be present in the F_2 generation in the proportion

A 1 in 16 **B** 3 in 16 **C** 6 in 16 **D** 9 in 16

5 In mice, black coat (allele B) is dominant to white coat (b) and straight whiskers (allele S) are dominant to curved whiskers (s).

When true-breeding mice with black coats and straight whiskers were crossed with white mice possessing curved whiskers, the offspring were all black with straight whiskers.

If these F_1 mice were crossed with white mice possessing curved whiskers, the expected proportion of offspring with black coats and curved whiskers in the next generation would be

A 1 in 16 **B** 3 in 16 **C** 4 in 16 **D** 9 in 16

6 When a true-breeding plant bearing disc-shaped yellow fruit was crossed with a plant bearing sphere-shaped green fruit, the F_1 generation were all found to bear disc-shaped yellow fruit.

This F_1 generation was self-pollinated and produced 800 offspring. Predict which of the following proportions of phenotypes was found in this F_2 generation.

	phenotype			
	disc-shaped yellow	*disc-shaped green*	*sphere-shaped yellow*	*sphere-shaped green*
A	199	201	202	198
B	49	153	152	446
C	800	0	0	0
D	453	148	147	52

7 In a certain species of sweet pea plant, flowers are either purple or white. Colour is determined by two unlinked genes. The alleles of the first gene are X and x; those of the second gene are Y and y.

In order to bear purple flowers, a plant must possess at least one X and one Y allele. Those genotypes which fail to do so, result in the formation of white flowers.

If two purple-flowered plants of genotype XxYy are crossed then the expected phenotypic ratio of offspring would be

A 12 purple : 4 white **B** 9 purple : 7 white
C 10 purple : 6 white **D** 8 purple : 8 white

8 In snapdragon plants, broad leaf is completely dominant to narrow leaf whereas red flower colour is incompletely dominant to ivory. (The genes for leaf width and flower colour are not linked.)

If a plant which is heterozygous for both genes is crossed with a true-breeding broad-leaved red-flowered plant, then the expected proportion of broad-leaved plants with pink flowers amongst the offspring would be

A 1 in 4. **B** 2 in 4. **C** 3 in 4. **D** 4 in 4.

9 The diagram opposite shows a pair of homologous chromosomes during meiosis.
Most crossing over will occur between genes

A W and X.
B X and Y.
C Y and Z.
D W and Z.

10 The diagram opposite shows a homologous (bivalent) pair of chromosomes during meiosis.
Which of the following correctly represents the final products of the second meiotic division?

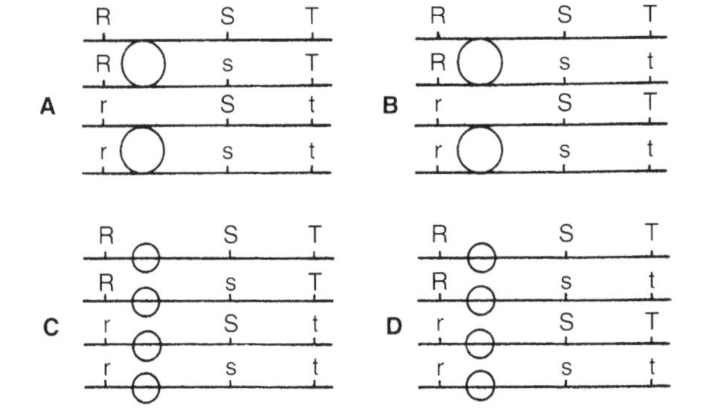

Items **11** and **12** refer to the following information.

In humans, the gene for red blood corpuscle shape (alleles elliptical E and normal e) is linked to the gene for Rhesus blood (alleles Rhesus positive R and Rhesus negative r).

11 A person with alleles E and R on one chromosome and e and r on its homologous partner will definitely produce gametes with the genotypes

A Ee and Rr. **B** Ee and er.
C ER and Rr. **D** ER and er.

12 If crossing over occurs between these two genes, then the two additional types of gametes that could result are

A RE and re. **B** EE and rr.
C Er and eR. **D** ee and RR.

13 The following crosses refer to experiments using the fruit fly, *Drosophila*.

True-breeding red-eyed flies with plain thoraxes were crossed with pink-eyed flies with striped thoraxes. The F_1 flies were then testcrossed against the double recessive as follows:

red eye × *pink eye*
plain thorax *striped thorax*

The following F_2 generation resulted from the cross.

80	16	12	92
red eye	*red eye*	*pink eye*	*pink eye*
plain thorax	*striped thorax*	*plain thorax*	*striped thorax*

What percentage number of recombinants resulted from the testcross?
A 12 **B** 14 **C** 16 **D** 28

14 In a certain species of animal, genes T, U, V and W occur on the same chromosome. The following table gives their cross-over values (COVs).

linked gene pair	COV
T and U	25
T and V	5
V and U	30
U and W	10
V and W	20

Which of the following represents the correct order of the genes on the chromosome?
A V, T, W, U. **B** T, W, U, V.
C T, V, W, U. **D** V, W, T, U.

15 J, K, L and M are four genes known to be located on the same chromosome. Testcrosses between heterozygotes and homozygous recessives are found to give the results shown in the following table.

cross	*% recombinants produced*
JjKk × jjkk	18
LlMm × llmm	38
MmJj × mmjj	24
LlKk × llkk	4

Which of the following diagrams best represents a linkage map of these genes on their chromosome?

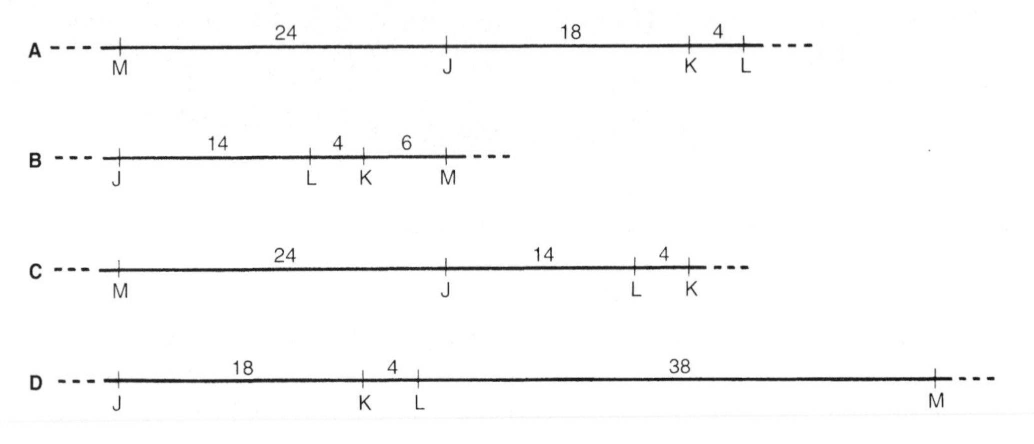

15

Sex linkage

Match the terms in list X with their descriptions in list Y.

list X		**list Y**	
1	autosome	**a**	inheritance pattern of alleles of genes carried on one sex chromosome but not on the other
2	carrier	**b**	sex-linked condition where blood fails to clot properly
3	haemophilia	**c**	smaller of the two human sex chromosomes
4	heterogametic	**d**	larger of the two human sex chromosomes
5	homogametic	**e**	gene present on X chromosome but not on Y chromosome
6	sex linkage	**f**	female who is heterozygous for a sex-linked gene
7	sex-linked gene	**g**	able to form one type of sex cell only
8	sex ratio	**h**	able to form two types of sex cell
9	X chromosome	**i**	type of chromosome not involved in sex determination
10	Y chromosome	**j**	relative number of males to females in a population

Choose the ONE correct answer to each of the following multiple choice questions.

1 Which of the following male animals is NOT heterogametic?

	animal	*chromosome complement*
A	fruit fly	2n = 6 + XY
B	fowl	2n = 14 + XX
C	grasshopper	2n = 16 + XO
D	human	2n = 44 + XY

Items 2 and 3 refer to the chromosome complement of each sex of fruit fly shown in the diagram.

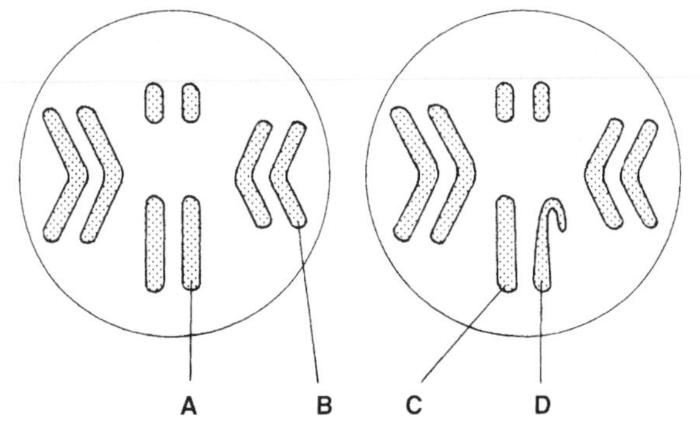

2 By which letter is an autosome labelled?

3 By which letter is a Y chromosome labelled?

4 A sex-linked allele NEVER passes from a
A man to his sons. **B** woman to her daughters.
C man to his grandsons. **D** woman to her granddaughters.

Items 5 and 6 refer to eye colour in the fruit fly. In this sex-linked trait, the allele for red eye is dominant to that for white eye.

5 If a red-eyed male is crossed with a white-eyed female, their offspring will occur in the ratio
A 1 red-eyed female : 1 red-eyed male.
B 1 white-eyed female : 1 red-eyed male.
C 1 red-eyed female : 1 white-eyed male.
D 1 red-eyed female : 1 white-eyed female : 1 red-eyed male : 1 white-eyed male.

6 If a heterozygous red-eyed female is crossed with a white-eyed male, what percentage of the female offspring will be white-eyed?
A 0% **B** 25% **C** 50% **D** 100%

7 Haemophilia is a condition in which blood fails to clot or clots only very slowly. Studies of this human sex-linked trait show that
A every X chromosome carries the dominant allele.
B a Y chromosome never carries the dominant allele.

C both X and Y chromosomes can bear the recessive allele.
D neither X nor Y chromosomes can bear the recessive allele.

8 Which of the following offspring could be produced by a normal homozygous female and a haemophiliac male?
A normal males and normal females
B haemophiliac males and normal females
C normal males and carrier females
D haemophiliac males and carrier females

9 A human female will definitely be a haemophiliac if
A both of her parents are also haemophiliacs.
B her mother is a carrier and her father is a haemophiliac.
C her mother carries the allele for haemophilia on both X chromosomes.
D her father is a haemophiliac and her mother is normal.

Items **10, 11** and **12** refer to the following information and family tree.

In humans, type of tooth enamel is a sex-linked trait. Brown tooth enamel (e) is recessive to normal (E).

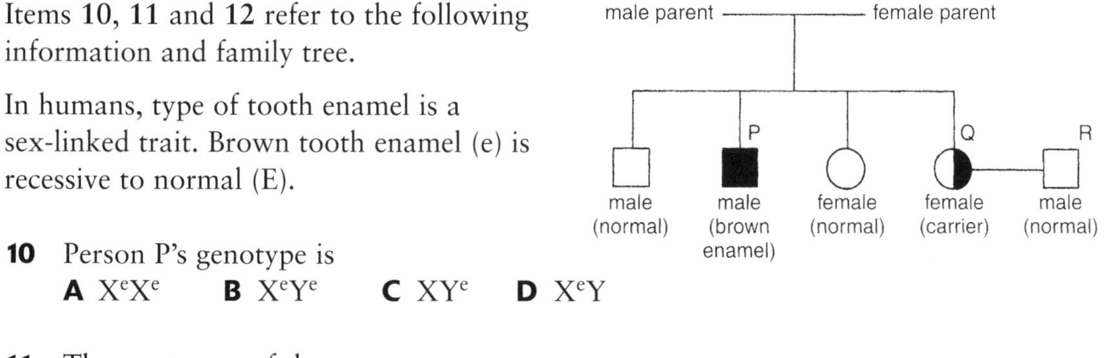

10 Person P's genotype is
A X^eX^e **B** X^eY^e **C** XY^e **D** X^eY

11 The genotypes of the two parents are
A X^EY and X^EX^e **B** X^eY and X^EX^E
C X^eY and X^EX^e **D** X^EY and X^eX^e

12 The probability of a child produced by persons Q and R being a boy with brown tooth enamel is
A 1 in 1. **B** 1 in 2. **C** 1 in 3. **D** 1 in 4.

13 In cattle, the male is heterogametic. When a normal male is crossed with a female heterozygous for a sex-linked **lethal** gene, the sex ratio of their living offspring will be

	female : male
A	3 : 1
B	2 : 1
C	1 : 1
D	1 : 2

14 Red-green colour-blindness is a sex-linked trait in humans. X^C = normal allele and X^c = colour-blind allele in the following cross.

$X^CX^c \times X^cY$

↓

X^cY

(Pat)

Which of the groups shown below consists of Pat's grandparents?

	maternal grandmother	*maternal grandfather*	*paternal grandmother*	*paternal grandfather*
A	X^CX^C	X^cY	X^CX^c	X^CY
B	X^CX^c	X^cY	X^CX^C	X^cY
C	X^CX^c	X^CY	X^CX^c	X^CY
D	X^CX^C	X^cY	X^CX^C	X^cY

15 In the magpie moth, wing colour is controlled by a sex-linked gene (where normal wing colour is dominant to pale colour).

In poultry (Light Sussex variety), plumage colour is controlled by a sex-linked gene (where white is dominant to red).

In the parental generation (P) of each of the following crosses, the homogametic sex is homozygous for the colour gene.

magpie moth

P pale male × normal female

↓

F_1 1 normal male : 1 pale female

poultry

P red male × white female

↓

F_1 1 white male : 1 red female

From these results it can be concluded that the heterogametic sex is

A male in magpie moth and female in poultry.
B female in magpie moth and male in poultry.
C male in both magpie moth and poultry.
D female in both magpie moth and poultry.

16 Mutation

Match the terms in list X with their descriptions in list Y.

list X		list Y	
1	deletion	**a**	doubling up of part of a chromosome involving several genes
2	Down's syndrome	**b**	loss of a segment of chromosome consisting of several genes
3	duplication	**c**	condition characterised by the chromosome complement 2n = 44 + XO
4	inversion	**d**	form of mutation involving one or more extra sets of chromosomes being added to a species' chromosome complement
5	karyotype	**e**	transfer of a segment of genes from one chromosome to another non-homologous one
6	Klinefelter's syndrome	**f**	increase in chromosome number of a cell caused by spindle failure during meiosis or mitosis
7	non-disjunction	**g**	reversal of the gene order of a segment of chromosome as a result of two breaks in the same chromosome
8	polyploidy	**h**	condition resulting from non-disjunction of chromosome 21 during meiosis
9	translocation	**i**	condition characterised by the chromosome complement 2n = 44 + XXY
10	Turner's syndrome	**j**	display of matched chromosomes showing their number, form and size

Match the terms in list X with their descriptions in list Y.

list X		list Y	
1	deletion	**a**	gene mutation involving the exchange of one nucleotide for another in the DNA chain
2	insertion	**b**	individual whose phenotype expresses a mutation
3	inversion	**c**	gene mutation involving the addition of an extra nucleotide to the DNA chain
4	mutagen	**d**	general term for a change in an organism's genetic material
5	mutant	**e**	gene mutation involving the loss of one nucleotide from the DNA chain
6	mutation	**f**	agent which increases mutation rate
7	substitution	**g**	gene mutation involving the reversal of the order of two or more nucleotides in the DNA chain.

Choose the ONE correct answer to each of the following multiple choice questions.

1 A mutation is a
A sudden temporary change in an organism's genetic material.
B change in phenotype followed by a change in genotype.
C change in hereditary material directed by a changing environment.
D change in genotype which may result in a new expression of a characteristic.

2 Which of the following statements is NOT correct?
A Mutations provide variation upon which natural selection can act.
B The vast majority of mutations produce alleles which are dominant.
C Mutations arise spontaneously, infrequently and at random.
D Mutation rate can be increased by artificial means.

3 A comparison of the karyotypes of a normal human male and a male sufferer of Down's syndrome shows the latter to possess
A one extra chromosome.
B two Y chromosomes.
C one extra pair of chromosomes.
D twice the normal number of chromosomes.

4 If a gamete mother cell of chromosome complement 44 + XY suffers a non-disjunction at the first meiotic division, which of the following sets of gametes could result?

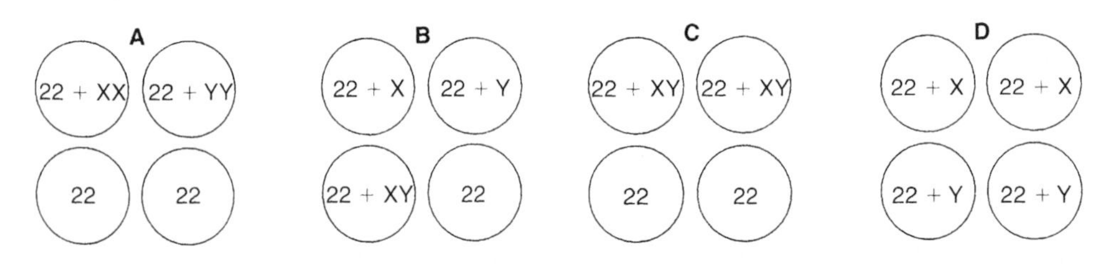

5 Klinefelter's syndrome results from the fusion of
A an X egg and a YY sperm.
B an XY egg and an X sperm.
C an XX egg and a Y sperm.
D an XX egg and a YY sperm.

6 Which chromosome complement is possessed by a sufferer of Turner's syndrome?
A 44 + XO
B 44 + XXY
C 45 + XX
D 45 + XY

7 A species of plant is known to have a diploid chromosome number of 14 in each of its cells. Which of the following would be the number of chromosomes found in each cell of one of its polyploid relatives?

A 7 **B** 14 **C** 15 **D** 28

8 Polyploid wheat does NOT normally show an increase in

A size. **B** vigour.
C resistance to disease. **D** length of life cycle.

9 The following diagram shows two chromosomes. The lettered regions represent genes.

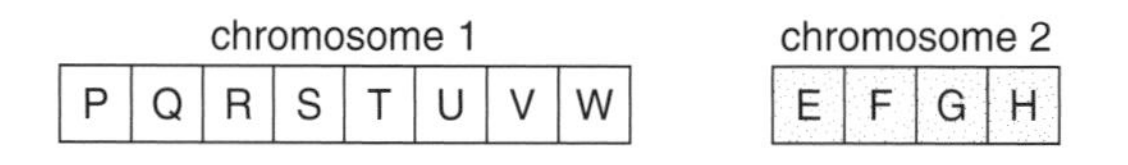

Which of the following would result if a translocation occurred between chromosomes 1 and 2?

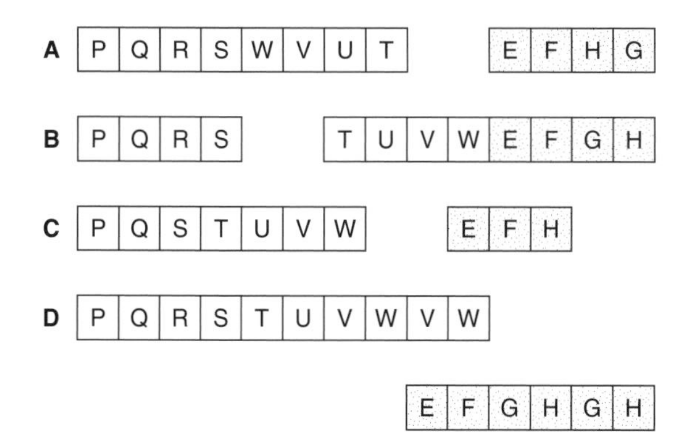

10 The following diagram shows the outcome of a certain type of chromosome mutation. The lettered regions indicate the positions of six marker genes.

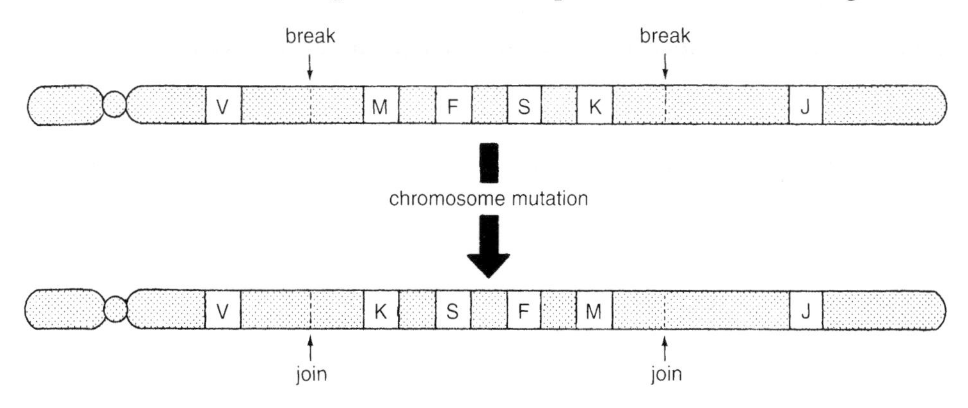

Which of the following diagrams best represents this mutated chromosome paired with its unaltered homologous partner during meiosis?

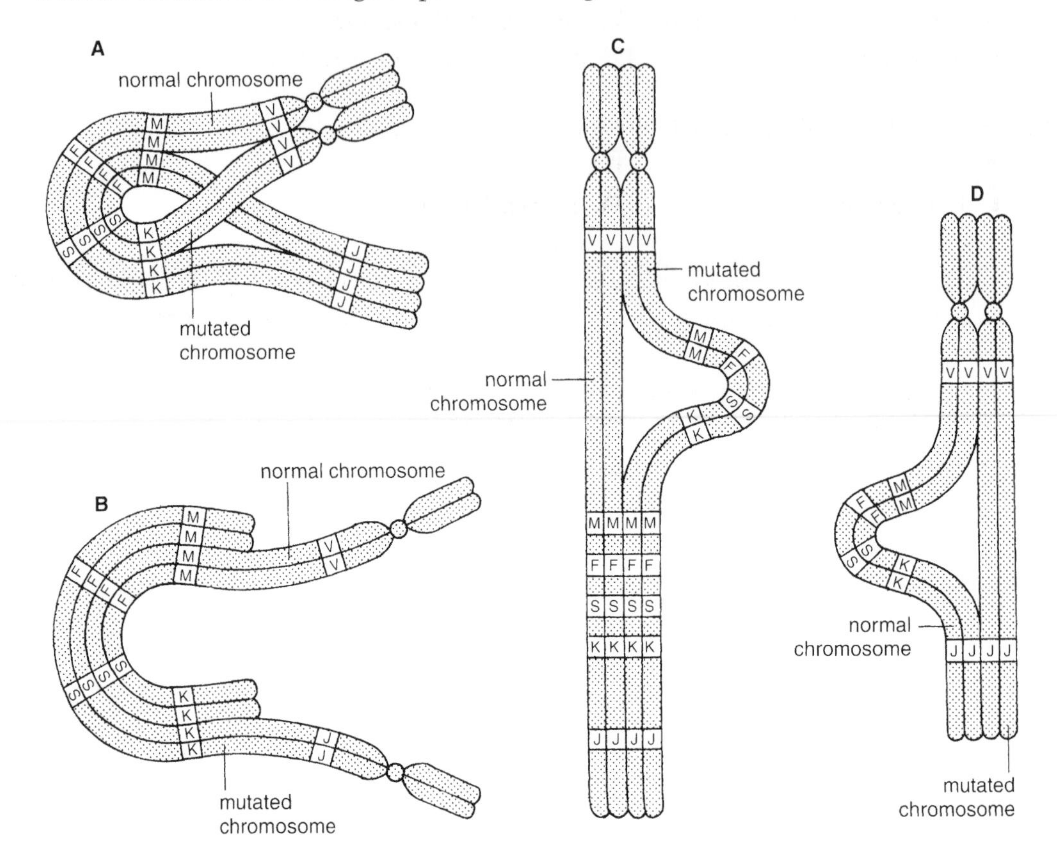

Items **11** and **12** refer to the following possible answers.

A inversion **B** deletion

C insertion **D** substitution

11 What name is given to the type of gene mutation where one incorrect nucleotide occurs in place of the correct nucleotide in a DNA chain?

12 What name is given to the type of gene mutation illustrated in the following diagram?

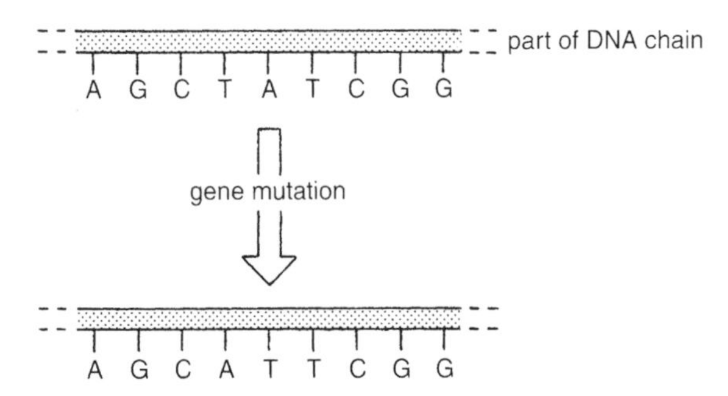

13 Which of the following gene mutations BOTH lead to a major change that causes a large portion of the gene's DNA to be misread?

A substitution and deletion
B deletion and insertion
C insertion and inversion
D inversion and substitution

14 Which of the following is NOT a mutagenic agent?

A nerve gas
B X-rays
C low temperature
D ultraviolet rays

15 Neurofibrosis, a condition in which the human sufferer develops multiple brown lumps in the skin, is caused by a dominant mutant allele whose mutation rate is 100 per million gametes.

The chance of a new mutation occurring is therefore

A 1 in 1000
B 1 in 10 000
C 1 in 100 000
D 1 in 1 000 000

17 Natural selection

Match the terms in list X with their descriptions in list Y.

list X

1. antibiotic
2. antibiotic-resistant
3. calcicole
4. calcifuge
5. conjugation
6. heavy metal tolerance
7. mutant allele
8. natural selection
9. selective advantage
10. sickle cell trait

list Y

- **a** process by which individuals best adapted to the environment survive and pass their genes on to succeeding generations
- **b** characteristic shown by some mutant plants enabling then to thrive in soil polluted with copper or lead
- **c** benefit gained by mutant organisms in an environment which suits them but not other members of the population
- **d** version of a gene containing genetic material different from the original
- **e** condition where the person's blood contains some haemoglobin S making him/her resistant to malaria
- **f** type of micro-organism whose growth is not inhibited by chemicals such as penicillin
- **g** member of a pair of plant ecotypes that grows best on acidic soils
- **h** member of a pair of plant ecotypes that grows best on calcium-rich soils
- **i** chemical made by one micro-organism that inhibits the growth of some other micro-organisms
- **j** form of contact between two bacteria which allows the transfer of genetic material from one to the other

Choose the ONE correct answer to each of the following multiple choice questions.

1 Which of the following does NOT comprise part of the theory of evolution proposed by Darwin in his book *The Origin of Species*?

A A struggle for existence occurs because organisms tend to produce more offspring than the environment will support.
B The members of the same species are not identical but show variation in all characteristics.
C Any beneficial change in an organism's phenotype is brought about by the direct action of the environment.
D Those offspring whose phenotypes are less well suited to the environment are likely to die before producing offspring.

Items **2**, **3**, **4**, **5**, **6** and **7** refer to the following information.

The peppered moth exists in two forms: the light-coloured variety and the dark (melanic) type. In an experiment, individuals of both types were marked on their underside with a dot of paint and then some were released in a rural area and some were released in an industrial area.

Many of these marked moths were later recaptured as shown in the following table.

	rural area		*industrial area*	
	light moth	*melanic moth*	*light moth*	*melanic moth*
number of marked moths released	250	200	250	see Item 4
number of marked moths recaptured	40	see Item 3	45	162
% number of marked moths recaptured	16	4	18	54

2 Melanic moths occur as a result of

A industrial pollution. **B** natural selection.
C speciation. **D** mutation.

3 How many melanic moths were recaptured in the rural area?

A 2 **B** 4 **C** 8 **D** 20

4 How many melanic moths were originally released in the industrial area?

A 200 **B** 250 **C** 300 **D** 350

5 From the data in the table it is NOT valid to conclude that

A in the rural area, light moths were four times more likely to survive than melanic moths.

B a greater % number of both types of moth were recaptured in the industrial area compared with the rural area.

C in the industrial area, melanic moths were three times more likely to survive than light-coloured moths.

D the total % number of light moths recaptured in both areas exceeded the total % number of melanic moths recaptured.

6 The reason for marking each moth on its UNDERSIDE was to

A treat all moths equally.

B make the paint inconspicuous to predators.

C avoid interfering with the moth's breathing system.

D prevent the paint from damaging the moth's wings.

7 The melanic moth enjoys a selective advantage in an industrial area because

A predators fail to notice it against a sooty background.

B there is no competition since the light form is killed by pollution.

C predators ignore it because it is dirty and noxious to eat.

D it is easily seen against light-coloured tree trunks.

8 The diagram opposite shows the outcome of a cross between two sufferers of sickle cell trait (where H = allele for normal haemoglobin and S = allele for haemoglobin S).

With respect to survival of the offspring, which of the following would be most likely?

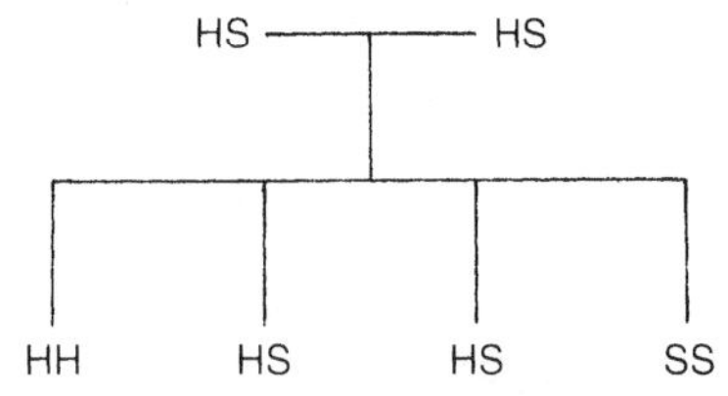

	% number of survivors	
	population living in malarial area	*population living in non malarial area*
A	25	75
B	50	75
C	25	100
D	50	100

9 The 'high speed' evolution of bacterial population 1 into bacterial population 2 shown in the accompanying diagram would involve THREE of the processes given in list Q. These are

A 1, 3 and 6.
B 1, 4 and 5.
C 2, 3 and 5.
D 2, 4 and 6.

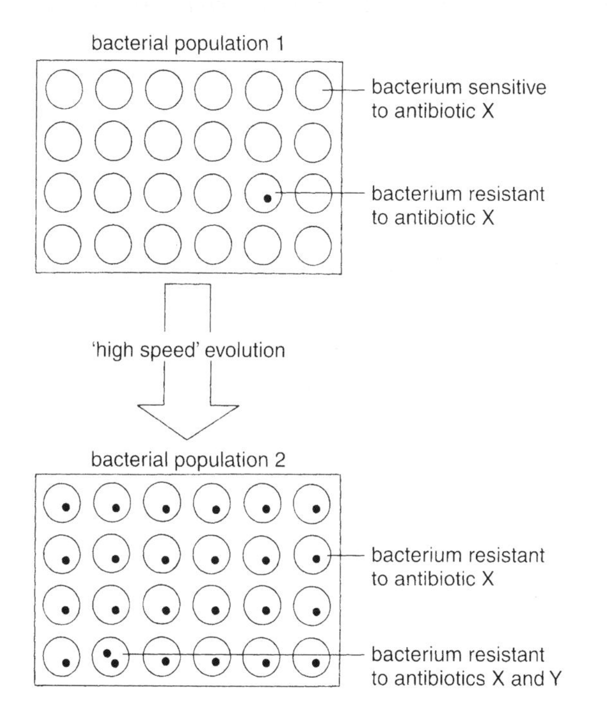

list Q

1 application of antibiotic X to population 1
2 application of antibiotic Y to population 2
3 multiplication of surviving bacterium resistant to antibiotic X
4 multiplication of surviving bacterium resistant to antibiotic Y
5 mutation producing bacterium resistant to antibiotic X
6 mutation producing bacterium resistant to antibiotic Y

10 Bacterial species P in the accompanying diagram is known to be resistant to the antibiotic streptomycin and to be able to transfer this characteristic to bacterial species R by conjugation.

Which letter indicates the possible location of the gene for resistance to streptomycin?

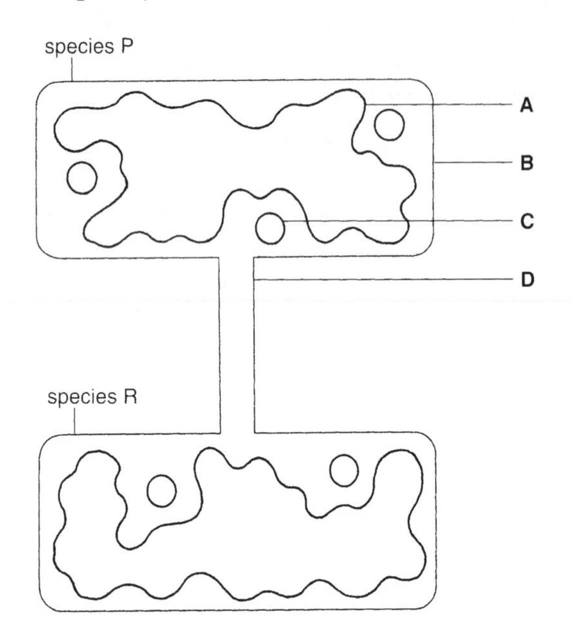

Items **11**, **12** and **13** refer to the following graph which shows the effect of an antibiotic on two strains of a species of bacterium.

11 To which concentration of antibiotic are both strains of the bacterium sensitive?

12 Which concentration of antibiotic affects neither strain of bacterium?

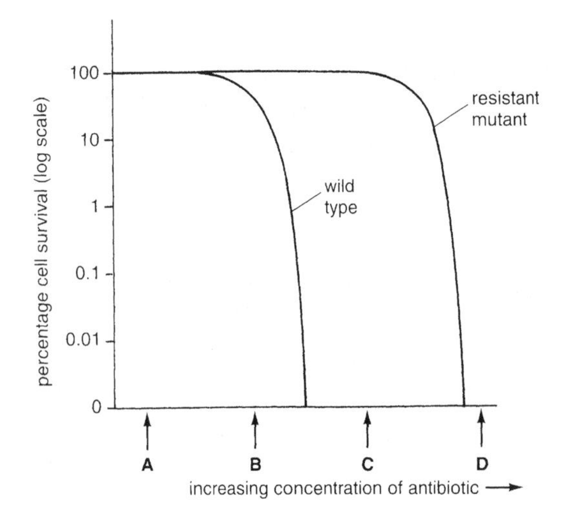

13 Which concentration of antibiotic would be suitable for use in selecting the resistant mutant strain?

14 Which of the following statements is FALSE? The use of antibiotics in the feed of farm animals

A prevents bacterial growth and promotes weight increase in farm animals.
B produces a strain of farm animals genetically resistant to disease.
C leads to the selection of bacteria resistant to antibiotics.
D increases the risk of resistant pathogens affecting humans.

15 The viral disease, myxomatosis, was deliberately introduced into Australia in the early 1950s in an attempt to control rabbit populations.

The following table shows the results from an investigation using rabbits selected each year from wild populations and inoculated with the original disease-causing strain of virus.

year	*% population suffering fatal symptoms*
1952	93
1953	95
1954	93
1955	61
1956	75
1957	54

These results support the theory that

A over the years an increased number of genetically resistant rabbits survived.
B natural selection occurred between 1955 and 1957 with a peak in 1956.
C the virus which caused myxomatosis underwent a mutation each year.
D rabbits acquired an immunity to the disease in 1956 only.

16 Certain species of grass are able to tolerate high concentrations of copper in the soil. An analysis was made of the grass plants at sites P, Q and R in the diagram opposite.

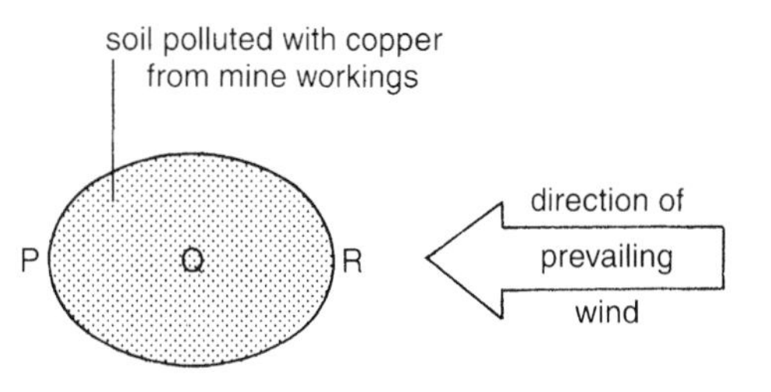

Which of the following correctly matches the three sites with the type(s) of grass plant that they would possess?

	all non-tolerant to copper	*some tolerant, some non-tolerant*	*all tolerant to copper*
A	P	Q	R
B	R	P	Q
C	R	Q	P
D	P	R	Q

17 Which of the following is NOT an example of natural selection in action?
A emergence of rats which thrive on warfarin rat poison
B development of pedigree strains of Rottweiler dogs
C resistance of certain types of bacteria to penicillin
D survival of mutant headlice treated with insecticide

Items 18, 19 and 20 refer to the following information and diagrams of land snails.

The shell of the land snail shows variation in both colour and banding pattern. In order to construct a 5-figure banding formula, bands are numbered from the top of the largest whorl as shown opposite. 0 is used to represent the absence of a band and square brackets indicate the fusion of two bands.

18 Shell S would have the banding formula
A 030[45] **B** 03045 **C** 02045 **D** 003[45]

19 Thrushes (which have good colour vision) smash the shells of land snails against stones (anvils) in order to feed on the soft inner body.

If snail types P, Q, R and S began in equal numbers in a habitat of grassland, which would be most likely to enjoy the greatest selective advantage?
A P **B** Q **C** R **D** S

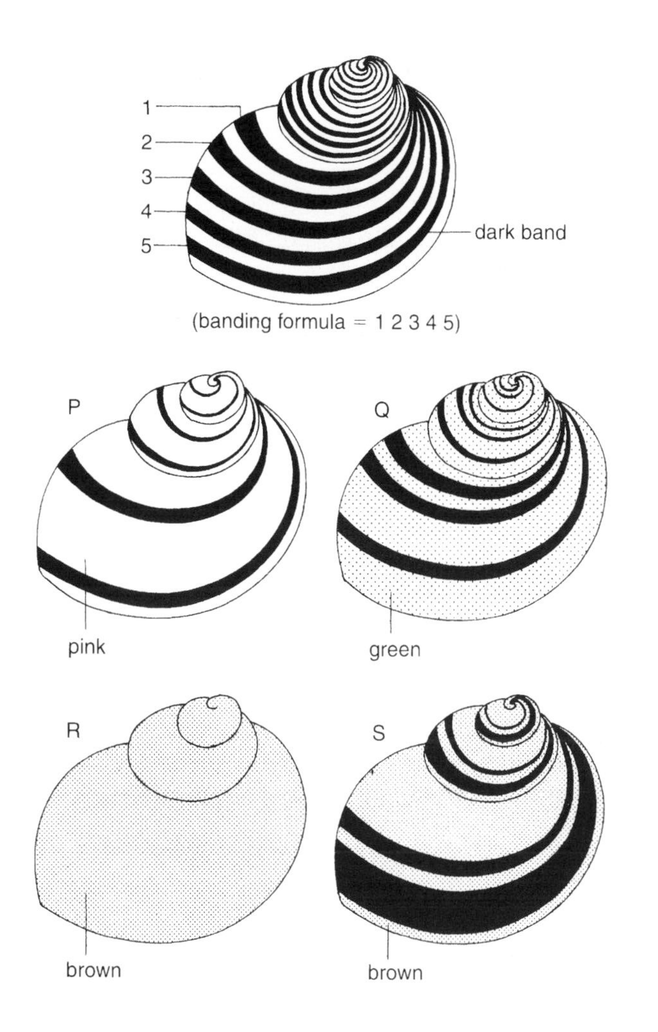

20 A survey of broken shells collected from thrush anvils amongst dead beech leaves in a woodland area was carried out. Predict which of the following sets of results was obtained.

	% number of broken shells of each type			
	P	*Q*	*R*	*S*
A	13	33	1	5
B	11	1	34	6
C	5	1	14	32
D	6	21	20	5

18

Speciation

Match the terms in list X with their descriptions in list Y.

list X	**list Y**
1 endemic	**a** total of all the different alleles present in the genotypes of the members of a population
2 continental drift	**b** barrier which prevents gene exchange between sub-populations of a species
3 gene frequency	**c** formation of new species
4 gene pool	**d** process by which the frequency of an allele increases if its phenotypic expression gives the organism an advantage
5 isolating mechanism	**e** group of organisms that produce fertile offspring and share the same basic anatomy and physiology
6 natural selection	**f** relative proportions of the alleles of a gene present in a population
7 speciation	**g** term used to describe organisms which only occur in a particular region such as an island
8 species	**h** process by which the world's land masses have gradually moved apart

Choose the ONE correct answer to each of the following multiple choice questions.

1 Which of the following occurs during the process of speciation?
A production of sterile offspring by interbreeding between two different species
B alteration in a species' phenotype caused by environmental change
C mass extinction of several species in a disturbed environment
D formation of new species from existing ones

2 The sum total of the genes possessed by the members of an interbreeding population at a given time is known as the
A gene frequency. **B** gene code.
C gene flow. **D** gene pool.

3 The members of the same species
A are reproductively isolated from one another.
B share different gene pools.
C possess the same chromosome complement.
D are unable to interbreed and produce fertile offspring.

4 Three of the events that occur during speciation are
1 mutation **2** natural selection **3** isolation
The correct order in which these occur is
A 3, 2, 1. **B** 2, 1, 3. **C** 3, 1, 2. **D** 2, 3, 1.

5 Which of the following acts as an ecological barrier during speciation?
A non-correspondence of genital organs
B occupation of habitats differing in altitude
C inability of sperm to fertilise eggs
D failure of insects to pollinate flowers

Items 6, 7, 8 and 9 refer to the following table which gives details of some Scottish varieties of *Apodemus*, the long-tailed field mouse. These animals live on the islands shown on the accompanying map on page 92.

scientific name			*location*	*average head and body length (mm)*	*average tail length (mm)*	*dorsal colour*
genus	*species*	*sub species*				
Apodemus	*hebridensis*	*hebridensis*	Lewis	95.5	87.8	wood brown
Apodemus	*hebridensis*	*tirae*	Tiree	102.5	84.2	reddish-brown
Apodemus	*hebridensis*	*hamiltoni*	Eigg	103.8	95.6	pale brown
Apodemus	*hebridensis*	*hirtensis*	St Kilda	110.9	105.5	peppery brown

6 How many species are shown in the table?
A 1 **B** 2 **C** 3 **D** 4

7 The isolating mechanism promoting speciation amongst the populations of *Apodemus* is
A altitude. **B** humidity.
C water. **D** temperature.

8 The type of isolating mechanism referred to in Item 7 is an example of
A a physiological barrier. **B** an ecological barrier.
C a reproductive barrier. **D** a geographical barrier.

9 A long-tailed Hebridean field mouse was found to have a head and body length of 101.9 mm and a tail length of 85.1 mm. It is most likely that it was a native of
A Lewis. **B** Tiree. **C** Eigg. **D** St Kilda.

10 Which of the following does NOT act as an ecological barrier during speciation?
A low temperature **B** high humidity
C acidic pH **D** mating behaviour

11 Areas X, Y and Z in the diagram at the top of page 93 represent three populations of a species of grassland-dwelling animal which is unable to fly or swim.

The barriers preventing interbreeding between populations X and Y and Y and Z respectively are
A desert and sea. **B** mountains and river.
C desert and river. **D** mountains and sea.

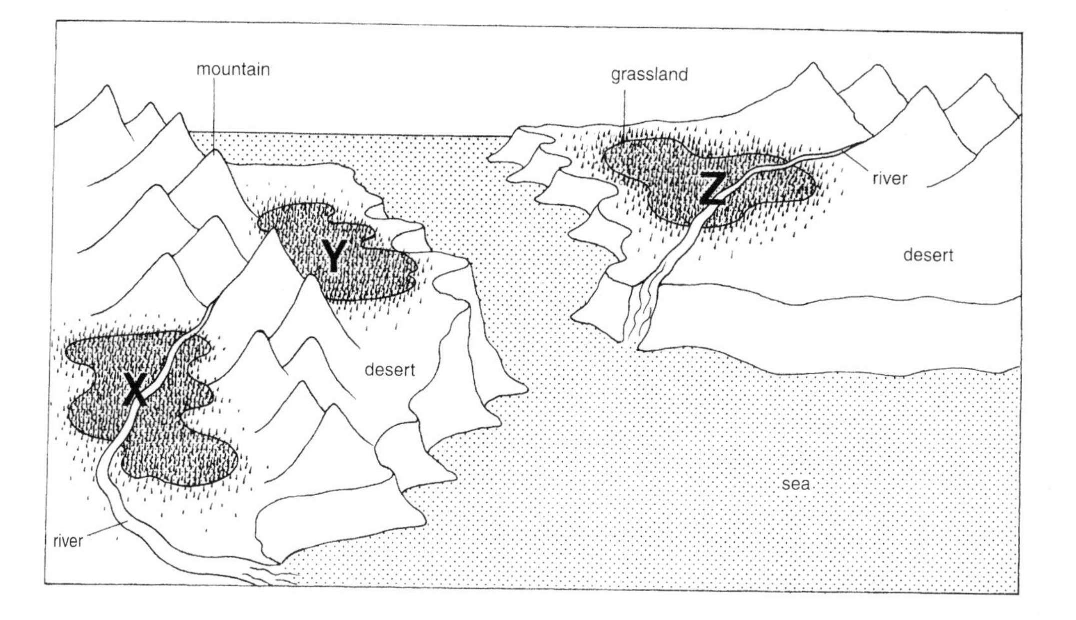

12 Which of the following factors can impose a selection pressure on a sub-population during the process of speciation?

A change in climate
B variation in genotype
C well adapted phenotype
D increased rate of mutation

13 *Sorbus arranensis* is a slender tree native to the island of Arran. Such a species that occurs only within a localised area is said to be

A biotic to the area.
B ecotypic of the area.
C endemic to the area.
D mutagenic of the area.

Items **14**, **15** and **16** refer to the following diagram which shows the geographical distribution of five populations of a certain type of sea bird.

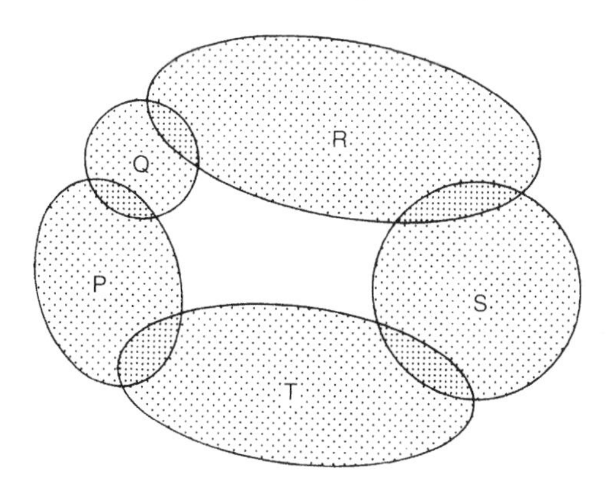

The table below shows the results of breeding experiments where + = successful interbreeding and – = unsuccessful interbreeding.

cross	*result*	*cross*	*result*
P × Q	+	Q × S	–
P × R	–	Q × T	–
P × S	–	R × S	+
P × T	–	R × T	–
Q × R	+	S × T	+

14 How many different species are present?
A 1 **B** 2 **C** 4 **D** 5

15 If population P became extinct, how many species would be present?
A 1 **B** 2 **C** 3 **D** 4

16 If, on the other hand, population S became extinct, how many species would be present?
A 1 **B** 2 **C** 3 **D** 4

17 The following diagram shows the positions that the world's land masses are thought to have occupied millions of years ago before they gradually drifted apart.

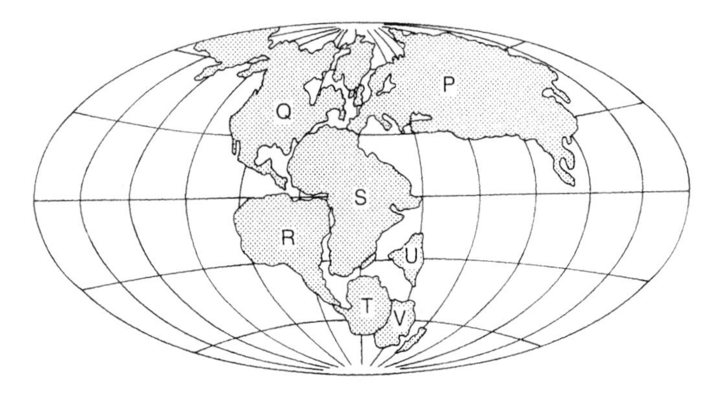

The spread of primitive mammals to Australia is thought to have taken the route
A S → R → T → V **B** R → S → T → V
C P → S → U → V **D** Q → S → U → V

18 There are three types of mammal:

1 *placentals* which retain their young in the uterus until a late stage of development;

2 *monotremes* which lay eggs and incubate them using body heat;

3 *marsupials* which bear their young at an early stage and rear them in a pouch.

Starting with the earliest and most primitive group, the order in which these evolved is:

A 2, 1, 3. **B** 2, 3, 1. **C** 3, 1, 2. **D** 3, 2, 1.

19 The land mass in the following map of the world is divided into six regions based on the types of mammal present in each region.

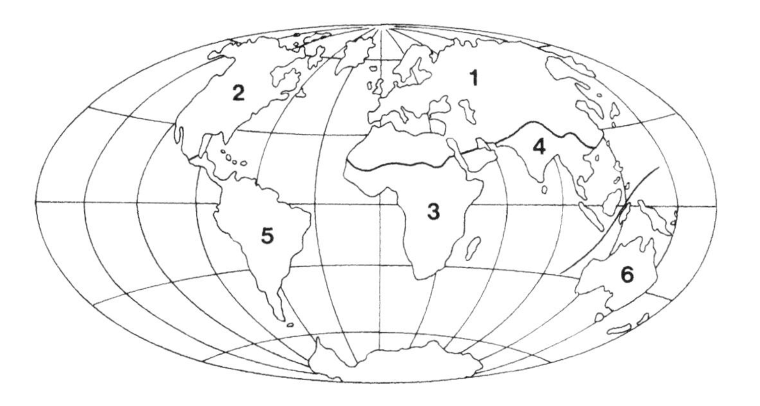

As a result of continental drift over millions of years, the mammals in

A regions 1 and 2 are very similar to one another.

B region 6 are very different to those in regions 1 and 2.

C regions 4 and 5 are very similar to one another.

D region 1 are very different to those in region 3.

20 The map on page 96 shows where some of the members of the camel family (Camelidae) are to be found.

Scientists believe that the camel family arose at

A 1 and migrated to 2, 3 and 4.

B 2 and migrated to 1, 3 and 4.

C 3 and migrated to 1, 2 and 4.

D 4 and migrated to 1, 2 and 3.

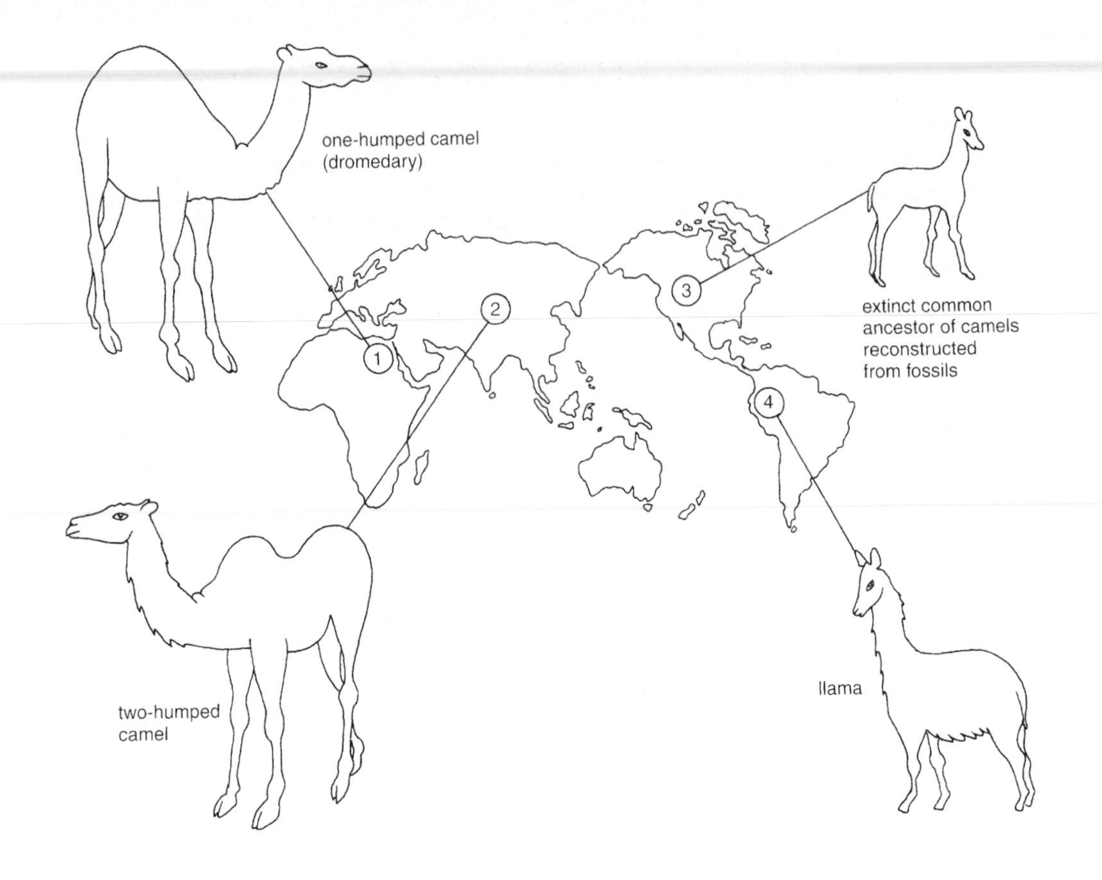
one-humped camel
(dromedary)
extinct common
ancestor of camels
reconstructed
from fossils
1
2
3
4
two-humped
camel
llama

19 Adaptive radiation

Match the terms in list X with their descriptions in list Y.

list X		list Y	
1	adaptive radiation	**a**	term describing the evolution of structures that differ in function but share a common ancestor
2	convergent	**b**	body parts of different species that share the same evolutionary origin but may differ in function
3	divergent	**c**	inherited characteristic that makes an organism well suited to its environment
4	ecological niche	**d**	mammal whose young are born after a relatively long gestation period in the mother's uterus
5	homologous structures	**e**	mammal whose young are born after a relatively short gestation period and reared in a pouch
6	marsupial	**f**	term describing the evolution of structures that perform the same function but do not share a close common ancestor
7	placental	**g**	process by which organisms evolve along separate lines involving adaptation to different environments
8	structural adaptation	**h**	role played by a species within a community

Choose the ONE correct answer to each of the following multiple choice questions.

1 Adaptive radiation is the evolution from
A a single ancestral stock, of several convergent forms adapted to share an available ecological niche.
B several divergent stocks, of a superior form adapted to fill an available ecological niche.
C a single ancestral stock, of several divergent forms adapted to fill different ecological niches.
D several divergent stocks, of several superior forms adapted to fill different ecological niches.

2 Scientists believe that over millions of years, Australian mammals have become very different from other mammals as a direct result of
A evolving pouches in which to rear their young.
B following their own course of evolution in isolation.
C developing reproductive systems homologous to placentals.
D evolving in climatically unique ecosystems.

3 Which of the following diagrams best represents the adaptive radiation of Darwin's finches?

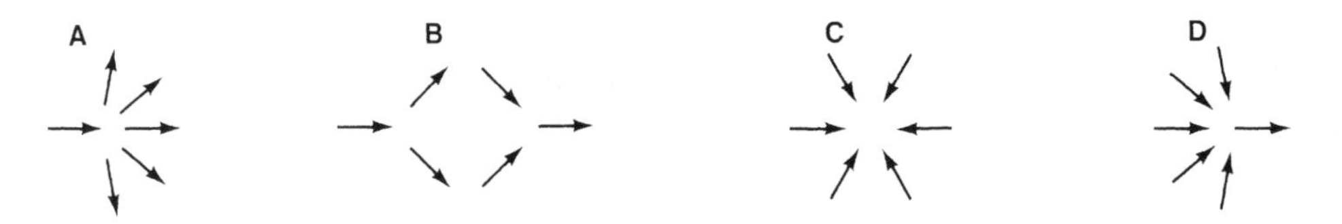

4 Which of the following statements is TRUE? Adaptive radiation enables each of the different species that evolves to
A exploit fully a particular set of environmental resources.
B compete effectively with all the others for the same resources.
C be a versatile feeder and make use of any available resource.
D make optimum use of the same ecological niche as the others.

5 The following list includes three Australian marsupial mammals and three placental mammals which occupy similar ecological niches in other continents of the world.

1 great red kangaroo
2 flying squirrel
3 sloth
4 phalanger
5 koala bear
6 deer

Which of the following correctly pairs each marsupial with the equivalent placental?
A 1 & 6, 4 & 3, 5 & 2
B 1 & 3, 4 & 2, 5 & 6
C 1 & 2, 4 & 3, 5 & 6
D 1 & 6, 4 & 2, 5 & 3

6 Which description in the following table refers accurately to homologous structures?

	basic structure	*common ancestor*	*function performed*
A	same	no	different
B	different	no	same
C	same	yes	different
D	different	yes	same

7 Which of the following show divergent evolution?
A eyes of locusts and blackbirds
B skeletons of tortoises and lobsters
C wings of cockroaches and pigeons
D forelimbs of bats and whales

8 Which of the following are examples of the evolution of homologous structures?
A legs of spider and crocodile
B canine teeth of wolf and gorilla
C hind limbs of tree frog and locust
D shells of snail and turtle

9 The diagram below shows the distribution of buttercup species on normal ridged pasture land.

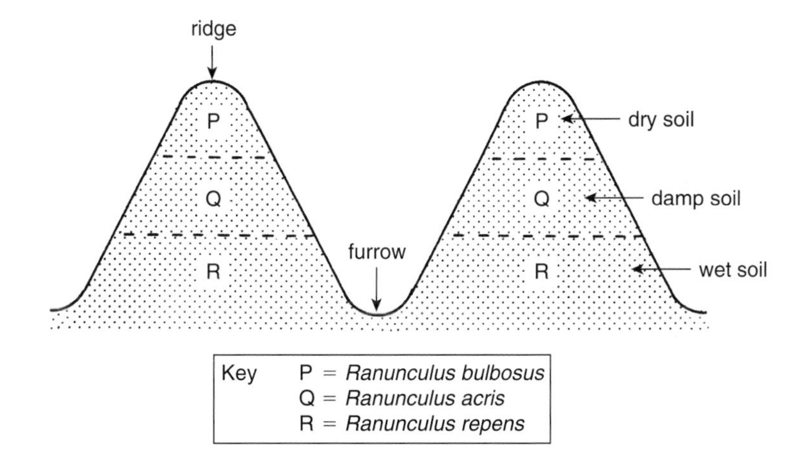

Which of the following shows the distribution of the buttercup species on ridged pasture land that is very poorly drained?

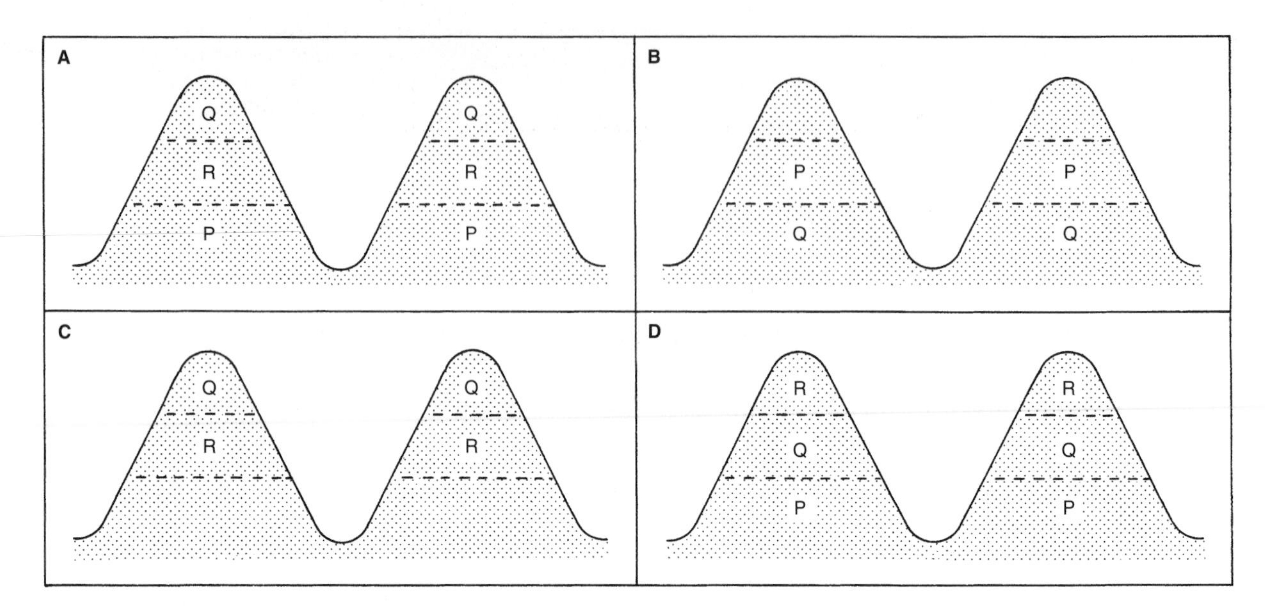

10

species	*habitat*			*food*		
	tree	*cactus*	*ground*	*seeds*	*buds and fruit*	*insects*
small ground finch			✓	✓		
cactus ground finch		✓	✓	✓		
large ground finch			✓	✓		
large cactus finch		✓		✓		
warbler finch	✓					✓
woodpecker finch	✓					✓
vegetarian tree finch	✓				✓	
insectivorous tree finch	✓					✓

The table refers to some of Darwin's finches. How many different ecological niches are being exploited by these eight species?

A 4 **B** 5 **C** 6 **D** 8

20 Extinction and conservation

Match the terms in list X with their descriptions in list Y.

list X		list Y	
1	cell bank	**a**	heritable variation that exists amongst living things
2	centre of diversity	**b**	irreversible loss of a species of living thing from planet Earth
3	conservation	**c**	types of living organism that thrive in habitats degraded by human activities
4	extinction	**d**	centre specialising in unusual or otherwise extinct relatives of farm animals
5	genetic diversity	**e**	region of the world possessing wild relatives of modern crop plants
6	habitat destruction	**f**	protection and careful management of a natural resource
7	opportunist species	**g**	human activity leading to possible extinction of many species
8	rare-breed farm	**h**	long-term storage site of seeds and sex cells of rare and valuable species

Choose the ONE correct answer to each of the following multiple choice questions.

1 About 440 million years ago the world was dominated by warm seas and much of the land mass was submerged. Five million years later, large ice sheets had formed, sea levels had dropped and vast areas of land had become exposed.

Which of the following types of organism would be most likely to suffer a wave of mass extinction under such circumstances?

A plankton living at the surface of the ocean
B deep sea invertebrate species
C marine animals adapted to warm shallows
D seaweeds native to cold waters

2 The process of extinction of species is decelerated by

A sudden climatic changes. **B** overhunting by humans.
C habitat destruction. **D** environmental conservation.

3 Which of the following animals was closest to worldwide extinction at the start of the 21st century?

A wild boar **B** giant panda
C European wolf **D** African elephant

4 At the beginning of the 21st century all of the following animals were in danger of becoming extinct EXCEPT

A red deer. **B** blue whale.
C grizzly bear. **D** Siberian tiger.

Items **5**, **6** and **7** refer to the following bar chart which shows eight crop plants which provided the world with 75% of its food during one year in the 1980s.

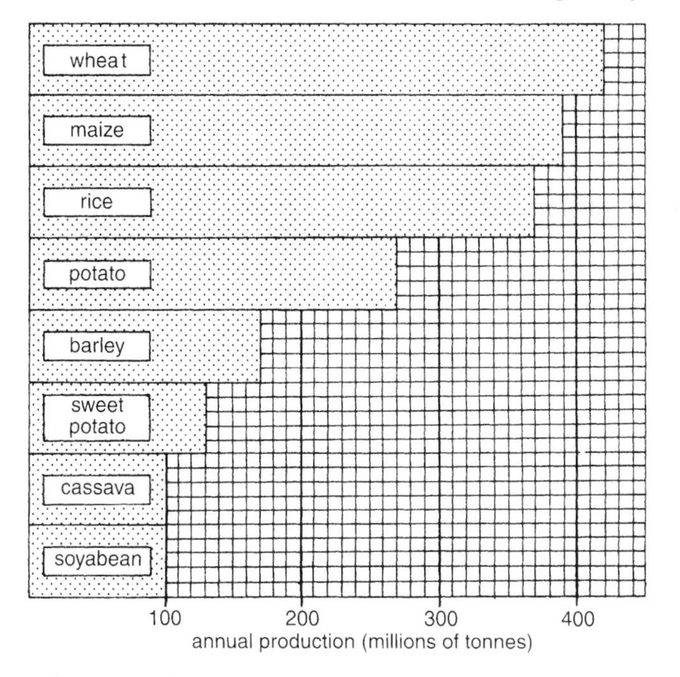

5 What was the TOTAL production of food (in millions of tonnes) obtained that year from ALL plant species in the world?
A 1462.5 **B** 1950.0 **C** 2500.0 **D** 2600.0

6 The percentage of the world's food production derived that year from sweet potato was
A 5.0% **B** 6.6% **C** 10.0% **D** 13.0%

7 Which plant ALONE provided the world with 15% of its needs?
A wheat **B** maize **C** rice **D** potato

8 About 11 regions of the world are known to still possess wild relatives of today's domesticated crop plants which are living in their natural habitats.

These regions are called
A cell banks. **B** botanical gardens.
C centres of diversity. **D** rare-breed farms.

9 It is important to conserve wild relatives of today's crop plants so that

A the wild plants' ecological niches can be used for crop growing in the future.

B alleles for disease resistance can be introduced into crops of the future.

C inferior alleles can be removed from crops and transferred to their wild relatives.

D nitrogen-fixing bacteria are prevented from dominating the soil ecosystem.

10 The conditions required for successful storage of live material in a cell (gene) bank are

	humidity level (%)	*temperature (°C)*
A	5	−20
B	5	20
C	95	−20
D	95	20

21 Artificial selection

Match the terms in list X with their descriptions in list Y.

list X		list Y	
1	cellulase	**a**	process by which only the organisms with the best features are chosen as the parents of the next generation
2	endonuclease	**b**	plant cell consisting of cell membrane, cytoplasm and nucleus but lacking a cell wall
3	gene probe	**c**	process by which protoplasts from two species form a hybrid containing a mixture of genetic traits
4	genetic engineering	**d**	enzyme used to seal a new gene into a bacterial plasmid
5	genome	**e**	process by which one variety of organism is crossed with a different variety to try to produce offspring better than either parent
6	hybrid vigour	**f**	short length of single-stranded DNA capable of binding to single-stranded DNA of required gene
7	hybridisation	**g**	enzyme used to cut DNA into fragments and open up bacterial plasmids
8	inbreeding	**h**	condition shown by offspring from cross between two different inbred parental strains
9	ligase	**i**	enzyme which digests cellulose in plant cell walls
10	protoplast	**j**	process by which genetic material from one organism is inserted into the genome of another
11	selective breeding	**k**	single haploid set of chromosomes typical of a species
12	somatic fusion	**l**	process by which close relatives are bred with one another and prevented from mating at random

Choose the ONE correct answer to each of the following multiple choice questions.

1 Which of the following diagrams best represents selective breeding in the *Brassica* group of plants?

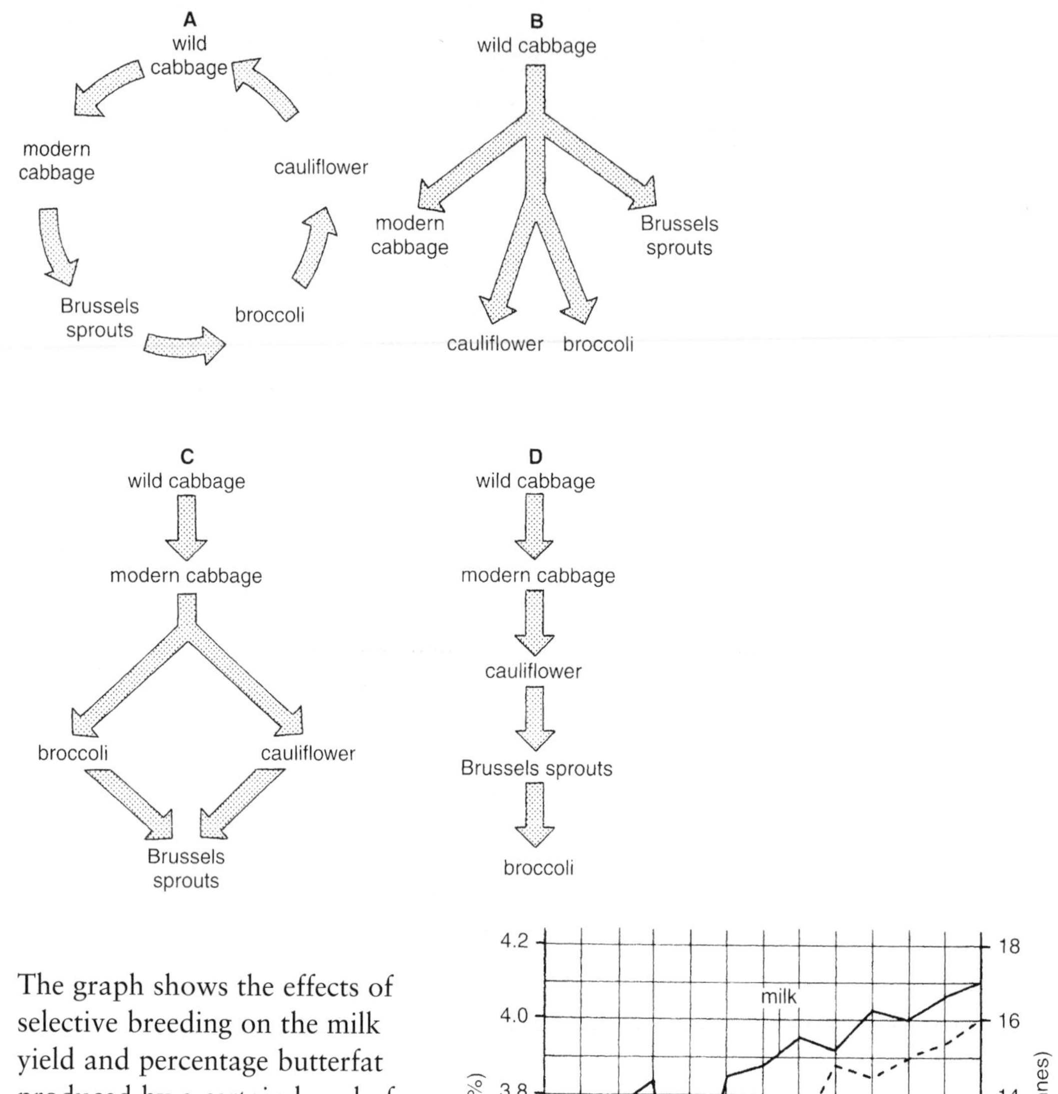

2 The graph shows the effects of selective breeding on the milk yield and percentage butterfat produced by a certain breed of cattle over a 30 year period.

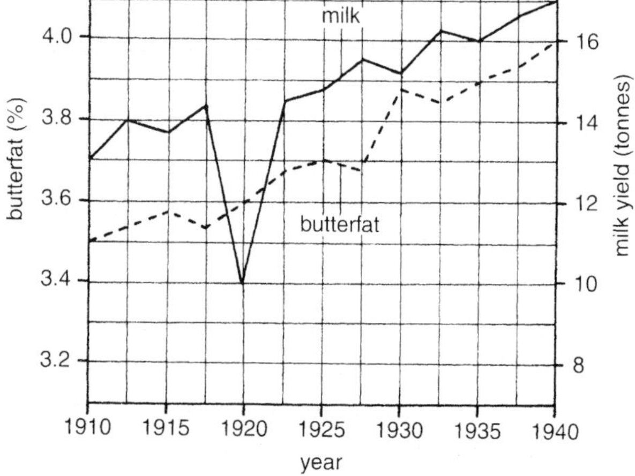

Which of the following shows the increases which occurred between 1910 and 1935?

	milk yield (tonnes)	*butterfat (%)*
A	0.4	3.0
B	0.5	4.0
C	3.0	0.4
D	4.0	0.5

3 Which of the following statements is FALSE?

A Inbreeding increases the chance of individuals arising that are double recessive for an inferior allele.

B Inbreeding depression often results from hybridisation between unrelated species.

C Inbreeding results in loss of genetic diversity amongst members of a domesticated variety.

D Inbreeding promotes the retention of desirable characteristics in a variety from generation to generation.

Items 4 and 5 refer to the following diagram which charts the effect of repeated self-pollination on heterozygosity in a variety of flowering plant.

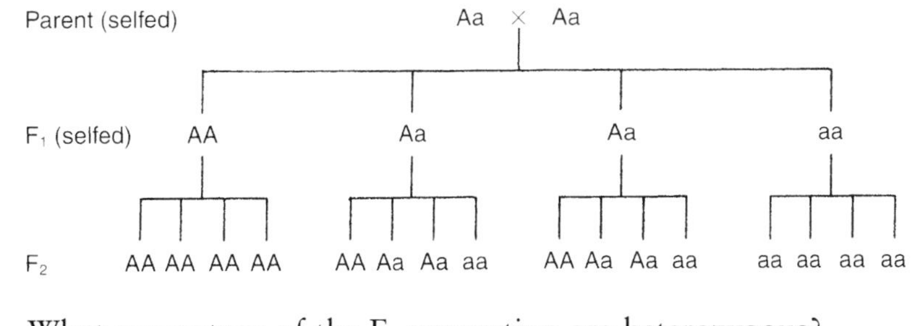

4 What percentage of the F_2 generation are heterozygous?

A 4 **B** 12 **C** 25 **D** 50

5 If the pattern of selfing were repeated, by which generation would there be less than 4% of heterozygotes remaining in the population?

A F_3 **B** F_4 **C** F_5 **D** F_6

6 Pedigree dogs are produced by mating members of the same breed with one another. This often results in the production of offspring suffering conditions which affect their fitness. For example, bulldogs have problems with their breathing and labradors are prone to arthritis.

This phenomenon is known as
A hybrid depression. **B** natural selection.
C inbreeding depression. **D** heterozygote selection.

7 Which of the following statements is FALSE?
A A hybrid is the result of a cross between genetically dissimilar parents.
B A hybrid is often stronger or better in some way than its parents.
C A hybrid formed from two different species is sterile because its chromosomes fail to pair at meiosis.
D A hybrid tends to be homozygous at many gene loci as a result of many generations of inbreeding.

8 In order to produce a supply of hybrids showing genetic uniformity, horticulturists often maintain two different true-breeding parental lines of a species of bedding plant.

The hybrids cannot be used as the parents of the next generation because
A a high mutation rate occurs amongst hybrid gametes.
B hybrids of annual plants always form sterile seeds.
C hybrid vigour cannot be passed on to the next generation.
D the hybrids are heterozygous and therefore not true-breeding.

9 Table 1 below shows the outcome of selfing four breeds of cattle (Q, R, S and T). Table 2 shows the outcome of hybridisation crosses involving the four breeds of cattle.

Table 1

parents	*average live weight of offspring at 18 months (kg)*
Q × Q	300
R × R	350
S × S	250
T × T	300

Table 2

parents	*average live weight of offspring at 18 months (kg)*
Q × R	320
R × S	310
Q × S	280
S × T	290

Which of the following crosses fails to show hybrid vigour since the offspring are poorer than the mean of the two parents?
A Q × R **B** R × S **C** Q × S **D** S × T

10 The following cross involves two varieties of the same species which have become homozygous as a result of many generations of inbreeding.

MMnnppQQ × mmNNPPqq

The offspring from this cross would have the genotype

A MMnnPPqq **B** MmNnPpQq

C mmNnPPQq **D** MMNNPPQQ

Items **11, 12** and **13** refer to the following information and the graphs that follow. Several types of artificial selection can be employed to alter a crop plant's quantitative characteristics.

1 DISRUPTIVE selection is practised when a crop is being developed for two different markets (e.g. barley with a low nitrogen content for brewing and barley with a high nitrogen content for livestock feed).

2 STABILISING selection is used for maintaining uniformity (e.g. crop height to suit harvesting machinery).

3 DIRECTIONAL selection is practised if increase in yield per plant is required.

11 Which of the sets of graphs shown below represents disruptive selection?

12 Which set of graphs represents stabilising selection?

13 Which set of graphs represents directional selection?

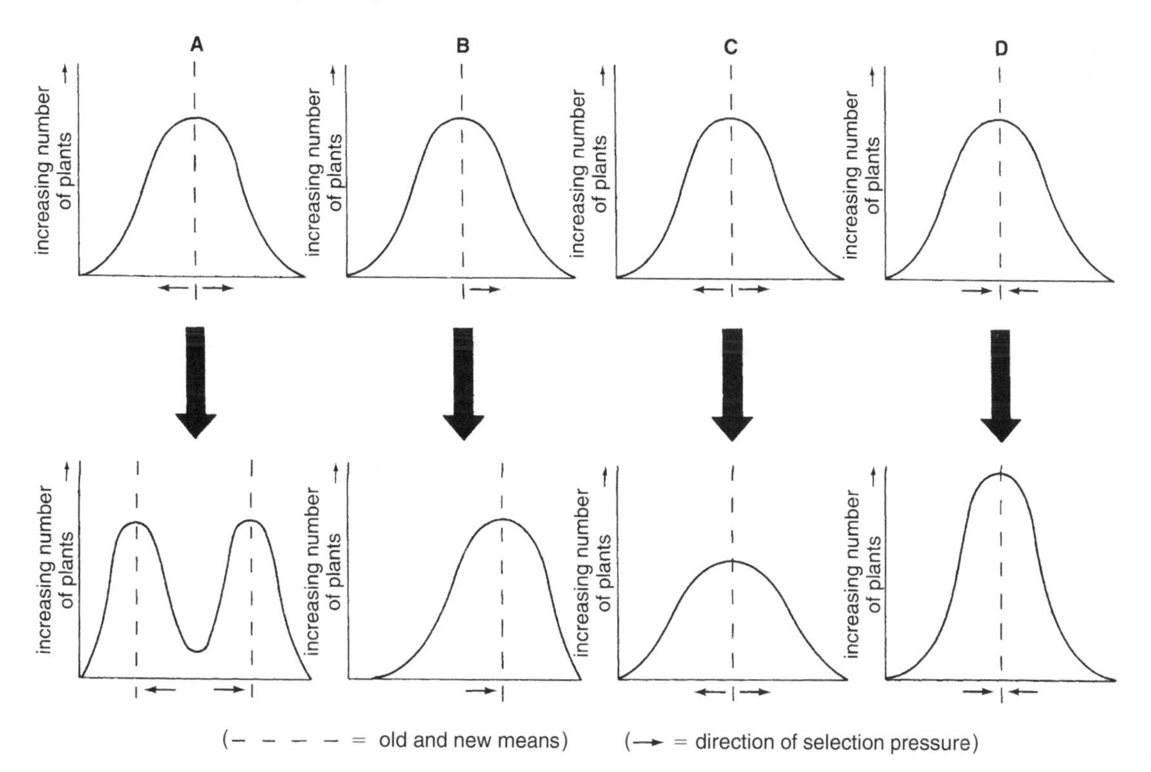

14 A gene probe can be detected easily because it

A contains thymine instead of uracil.
B carries a radioactive chemical label.
C comprises a unique series of anticodons.
D consists of a single strand of nucleic acid.

15 The central part of the diagram below shows a gene probe that has been produced to locate the gene for sickle cell haemoglobin (HbS) in humans.

The diagrams surrounding it show DNA fragments from the genetic material of four newborn babies. Which baby possesses the gene for sickle cell haemoglobin?

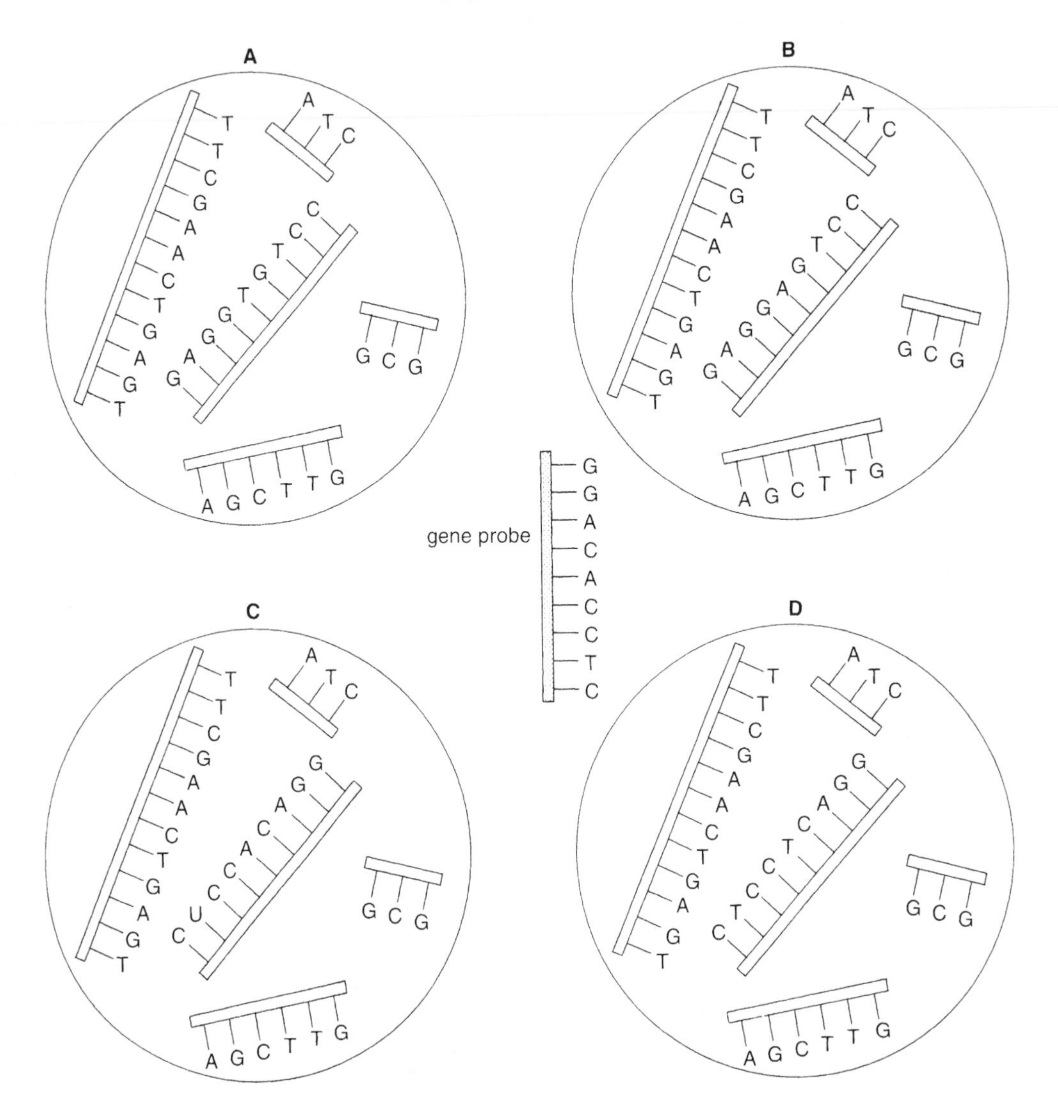

Items **16** and **17** refer to the following information.

A certain strain of bacterium *Streptomyces* possesses a gene that makes it resistant to a type of weedkiller. In order to produce a strain of potato also resistant to the weedkiller, genetic engineers have transferred the useful gene using as the vector a plasmid from a second type of bacterium that induces tumours in potatoes.

The following list gives the steps adopted in their procedure in the wrong order.

1 Gene resistant to weedkiller inserted and sealed into plasmid.
2 Gene resistant to weedkiller cut out of *Streptomyces* bacterium.
3 Samples of infected potato cells grown into 'transformed' plants resistant to weedkiller.
4 Plasmid extracted from tumour-inducing bacterium and opened up.
5 Plasmid returned to tumour-inducing bacterium which is allowed to infect potato cells.

16 Which answer indicates the correct sequence of steps?
A 2, 1, 5, 4, 3. **B** 4, 1, 2, 3, 5. **C** 2, 4, 1, 5, 3. **D** 4, 5, 2, 3, 1.

17 The enzyme ligase would be employed at
A step 1 only. **B** steps 1 and 2. **C** step 2 only. **D** steps 2 and 4.

Items **18** and **19** refer to the following information.

Restriction endonuclease enzymes do not cut DNA at random but recognise particular sequences of bases.

18 One enzyme has the following recognition sequence:

cut
↓
– – –G G A T C C– – –
– – –C C T A G G– – –
↑
cut

The diagram below shows a piece of DNA about to be cut.

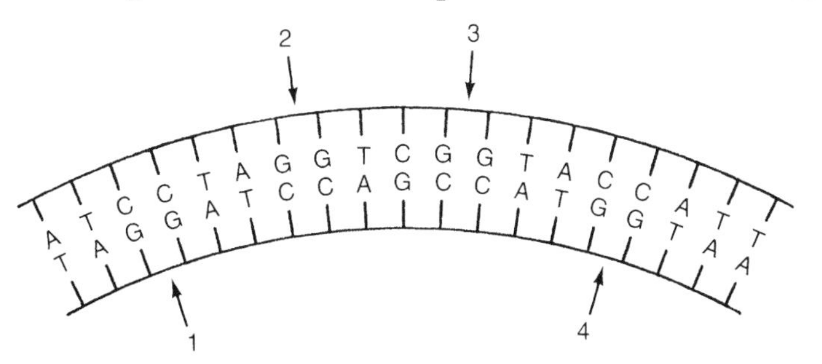

At which pair of numbered sites would the enzyme make its cuts?
A 1 and 2 **B** 3 and 4 **C** 1 and 3 **D** 2 and 4

19 The diagram below shows a different piece of DNA about to be acted upon by a second enzyme with the recognition sequence:

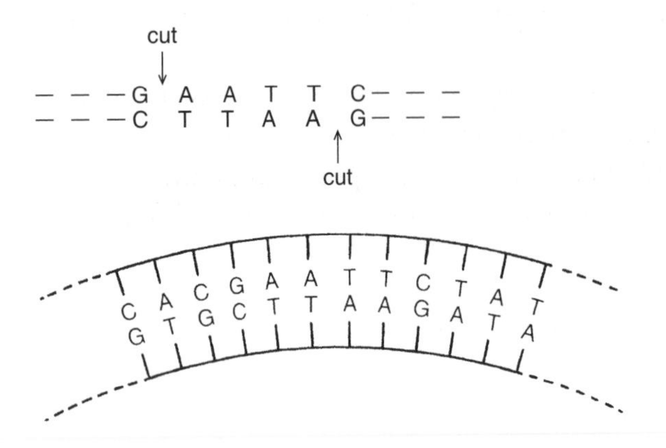

Which of the following diagrams shows the outcome of this enzyme's action?

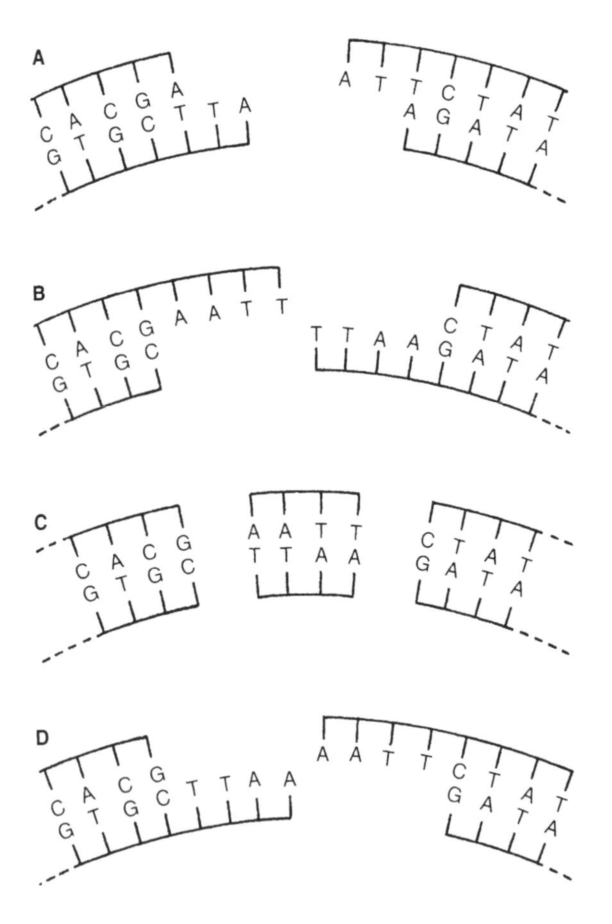

20 The list below gives the steps involved in bringing about somatic fusion between two species.

1 Isolated protoplasts are induced to fuse by employing chemicals or electric currents.

2 The callus is treated with hormones to make it develop into a hybrid plant.

3 Cell walls are removed using the enzymes cellulase and pectinase.

4 Unspecialised cells are selected from two different species of plant.

5 The hybrid protoplast is allowed to divide forming a mass of undifferentiated cells.

The correct sequence in which these steps would be carried out is

A 3, 4, 1, 5, 2.
B 4, 3, 5, 1, 2.
C 3, 4, 1, 2, 5.
D 4, 3, 1, 5, 2.

22 Maintaining a water balance – animals

Match the terms in list X with their descriptions in list Y.

list X		list Y	
1	adaptation	**a**	lower in water concentration compared with another solution
2	ammonia	**b**	control of water balance by a living organism
3	chloride secretory cell	**c**	process by which water passes from a hypotonic to a hypertonic solution
4	glomerulus	**d**	inherited characteristic which suits an organism to its environment and increases its chance of survival
5	hypertonic	**e**	equal in water concentration to another solution
6	hypotonic	**f**	nitrogenous waste made by freshwater fish
7	isotonic	**g**	nitrogenous waste made by saltwater fish
8	migration	**h**	nitrogenous waste made by mammals
9	osmoregulation	**i**	higher in water concentration compared with another solution
10	osmosis	**j**	site of active transport of salt ions into a freshwater fish and out of a saltwater fish
11	trimethylamine oxide	**k**	site of filtration of blood in a kidney
12	urea	**l**	behavioural adaptation which enables an animal to spend part of its life in one environment and part in another

Choose the ONE correct answer to each of the following multiple choice questions.

Items **1** and **2** refer to the following possible answers.

A jellyfish **B** haddock
C *Paramecium* **D** stickleback

1 Which animal possesses body contents which are hypotonic to the animal's natural surroundings?

2 Which animal's body contents are isotonic to its natural surroundings?

3 Which of the osmoregulatory mechanisms shown in the following diagram is employed by a freshwater bony fish?

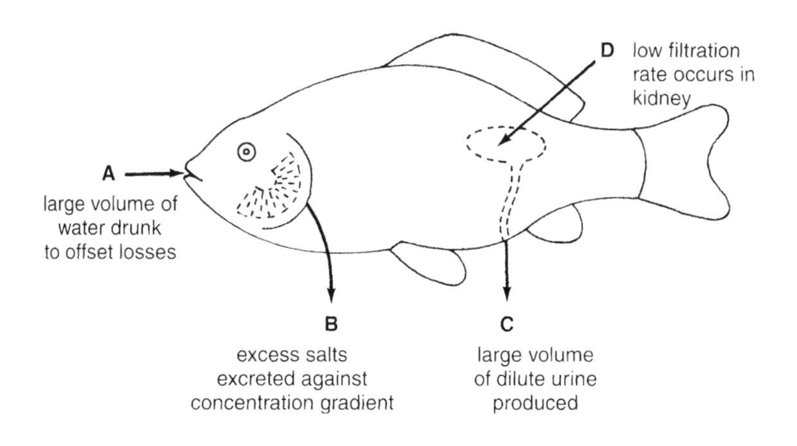

4 Which of the following pairs of characteristics would be true of a mammal which produces very concentrated urine?

	level of ADH in blood	*relative length of loops of Henle*
A	low	short
B	low	long
C	high	short
D	high	long

Items 5 and 6 refer to the possible answers in the following table.

	relative volume of water drunk	*relative volume of urine produced*
A	none	little
B	little	much
C	much	much
D	much	little

5 Which animal is a tuna fish?

6 Which animal is a desert rat?

7 The gills of a saltwater fish
A lose water by osmosis and absorb salts.
B gain water by osmosis and absorb salts.
C lose water by osmosis and excrete salts.
D gain water by osmosis and excrete salts.

8 The structures shown in the accompanying diagram are present in the kidney of a certain animal.

This animal could be a
A camel.
B desert rat.
C freshwater bony fish.
D saltwater bony fish.

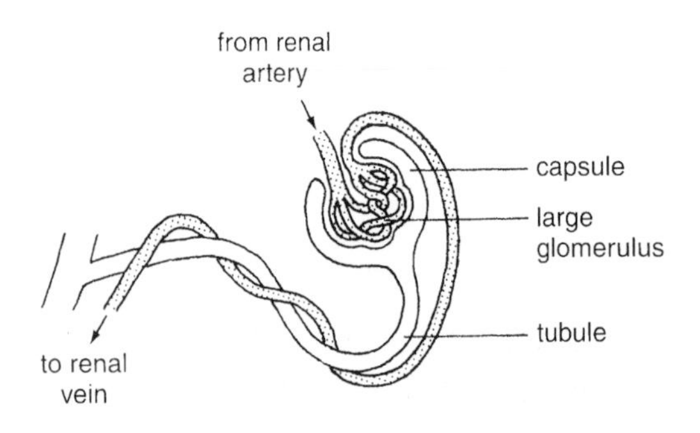

Items **9** and **10** refer to the following graph which charts the variation in salt concentration of the body fluid of four invertebrate animals when placed in different dilutions of sea water.

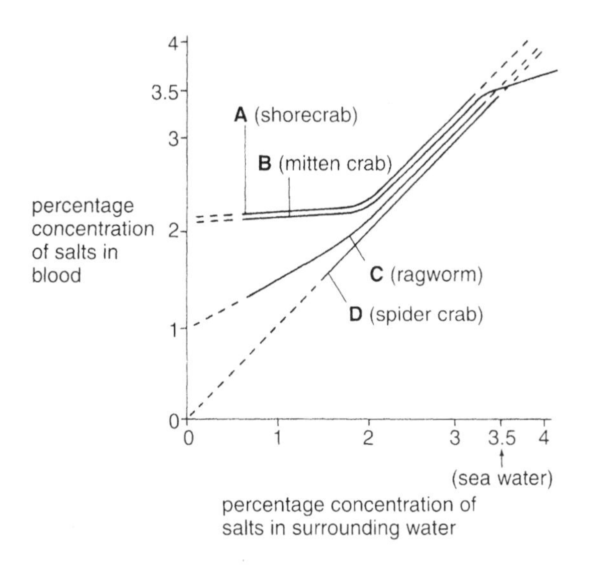

9 Which animal is completely unable to maintain a blood concentration of salt above that in the surrounding water?

10 Which animal is able to maintain a blood concentration of salt lower than that in the surrounding water?

11 The four routes by which water is lost from a horse are shown in the diagram. Which of these does NOT occur in a kangaroo rat?

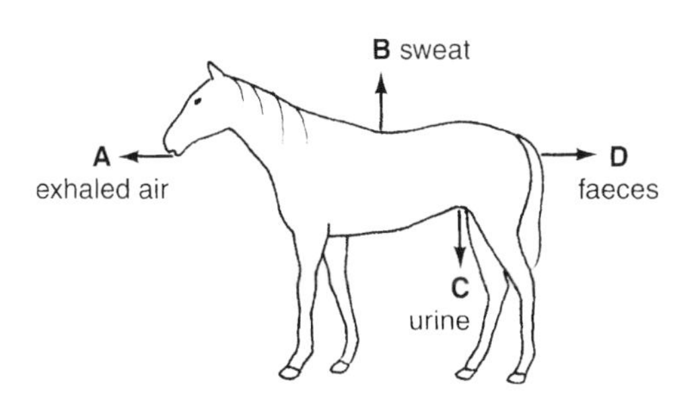

12 The following statements refer to the camel. Which one is FALSE?
A Its red blood cells are very sensitive to osmotic changes of the blood.
B It can drink 100 litres of water in a very short space of time.
C Its rate of urine production is low and little water is lost in faeces.
D It does not begin to sweat until its body temperature reaches 41°C.

13 Which of the following does NOT occur when an Atlantic salmon migrates from fresh water to salt water?

A change in filtration rate of kidneys
B extension in length of kidney tubules
C reversal in direction of salt transfer by gills
D reduction in volume of urine produced

Items **14** and **15** refer to the following graphs which summarise the results from a series of experiments on seawater-adapted and freshwater-adapted trout treated with a hormone called angiotensin II.

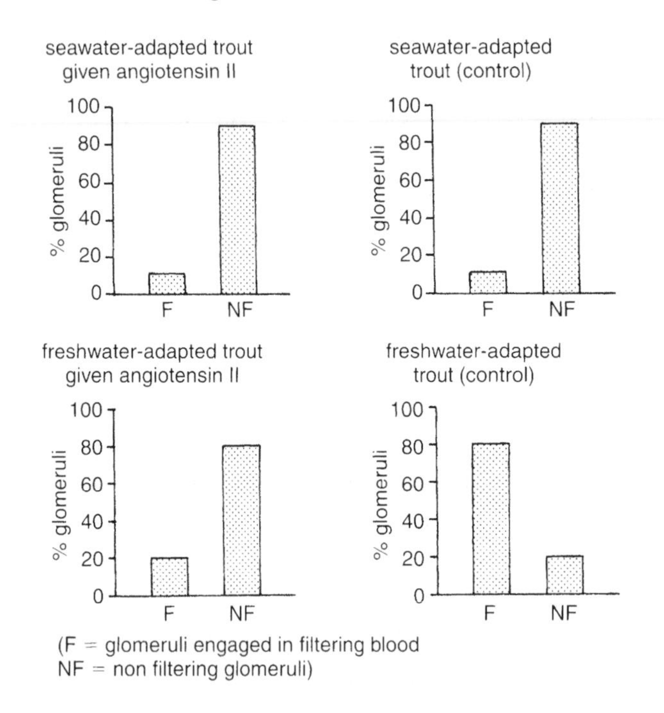

14 From this experiment it can be concluded that an increase in concentration of angiotensin II results in

A a reduction in number of filtering glomeruli in saltwater-adapted trout.
B an increase in number of filtering glomeruli in saltwater-adapted trout.
C a reduction in number of filtering glomeruli in freshwater-adapted trout.
D an increase in number of filtering glomeruli in freshwater-adapted trout.

15 In some fish, angiotensin II induces an antidiuretic effect thereby promoting water retention. This effect could be brought about by

A narrowing of renal arterioles which reduces rate of blood flow to glomeruli.
B widening of renal arterioles which reduces rate of blood flow to glomeruli.
C narrowing of renal arterioles which increases rate of blood flow to glomeruli.
D widening of renal arterioles which increases rate of blood flow to glomeruli.

23

Maintaining a water balance – plants

Match the terms in list X with their descriptions in list Y.

list X		list Y	
1	atmometer	**a**	the continuous passage of water through a plant from its roots to its leaves
2	capillarity	**b**	instrument used to measure the rate of water uptake by a plant
3	hydrophyte	**c**	force caused by the flow of water into a plant from the soil solution by osmosis
4	mesophyte	**d**	major force responsible for the ascent of water in the transpiration stream
5	osmoregulation	**e**	pore in a leaf surface controlled by guard cells
6	potometer	**f**	force of attraction between particles which makes water rise up narrow tubes
7	root pressure	**g**	loss of water by evaporation from the aerial parts of a plant
8	stoma	**h**	plant adapted to life completely or partially submerged in water
9	transpiration	**i**	plant adapted to a habitat where soil water is lacking or conditions are excessively windy
10	transpiration pull	**j**	'normal' plant adapted to a habitat where soil water is plentiful
11	transpiration stream	**k**	instrument used to measure the rate of evaporation of water from a non living surface
12	xerophyte	**l**	control of water balance by a living organism

Choose the ONE correct answer to each of the following multiple choice questions.

1 The ascent of sap in a plant is thought to be assisted by
1 transpiration pull **2** root pressure **3** capillarity.

Which of the above involve osmosis?
A 1 and 2 only **B** 1 and 3 only
C 2 and 3 only **D** 1, 2 and 3

2 The column of sap being pulled up a tree is maintained without a break because the water molecules
A adhere to one another.
B cohere to one another.
C adhere to the surrounding air molecules.
D cohere to the sides of the xylem vessels.

Items 3 and 4 refer to the following diagram of part of a transverse section of a leaf.

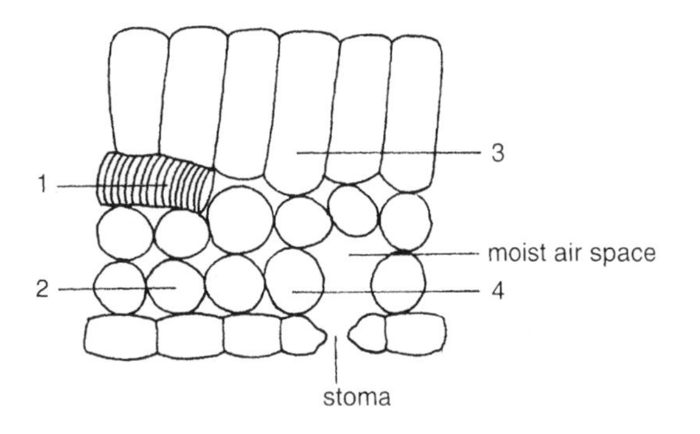

3 At which location is the water concentration highest?
A 1 **B** 2 **C** 3 **D** 4

4 For the transpiration pull to be effective, a water concentration gradient must exist between
A 1 and 3. **B** 1 and 4. **C** 2 and 3. **D** 2 and 4.

5 During a period of very rapid transpiration, the diameter of a tree trunk often
A increases because less water is present in the xylem vessels.
B decreases because more water is present in the xylem vessels.
C increases because tension decreases in the xylem vessels.
D decreases because tension increases in the xylem vessels.

Items **6** and **7** refer to the following graph.

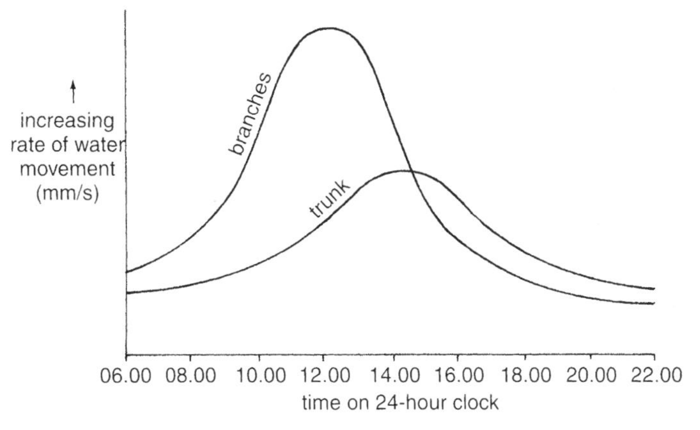

6 From the graph it can be correctly concluded that during the course of one day

A the increase in rate of water movement first begins in the trunk.
B the greatest increase in rate of water flow first occurs in the trunk.
C any change that first occurs in the trunk is repeated later in the branches.
D the greatest increase in rate of water flow first occurs in the branches.

7 The results presented in this graph support the theory that

A the greatest rate of transpiration normally occurs at 14.00 hours.
B the rate of water movement in the branches is not related to that in the trunk.
C the leaves provide the driving force for the upward movement of water.
D a larger volume of water passes through the branches than the trunk.

8 A stomatal pore opens as a result of the guard cells

A absorbing water by osmosis.
B using up sugar during respiration.
C using up water during photosynthesis.
D absorbing carbon dioxide by diffusion.

9 The following graph shows the rates of water absorption and transpiration that occurred in a sunflower plant during a 24-hour period.

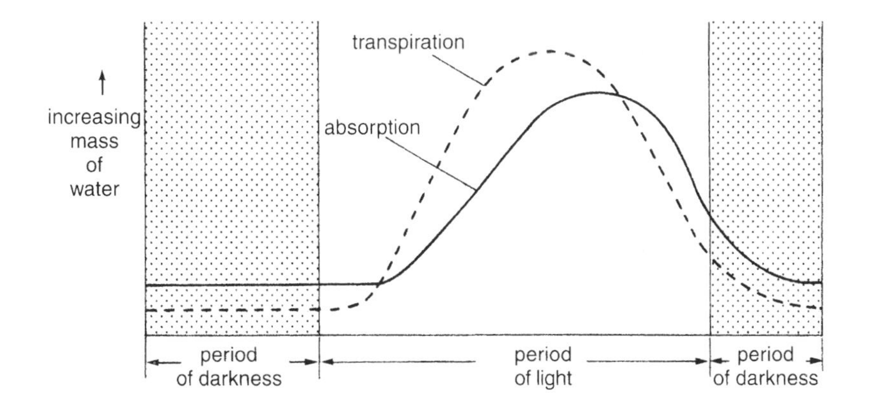

From this evidence it is true to say that in
A light, absorption rate always exceeds transpiration rate.
B dark, absorption rate always exceeds transpiration rate.
C light, transpiration rate always exceeds absorption rate.
D dark, transpiration rate always exceeds absorption rate.

10 The following diagram shows four different pieces of apparatus.

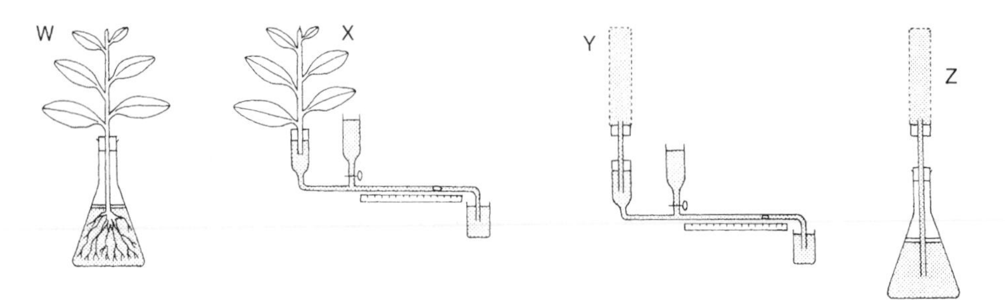

Which of these are BOTH atmometers?
A W and X **B** X and Y
C W and Z **D** Y and Z

11 Which of the following is NOT an essential precaution taken when setting up and using a bubble potometer?
A cutting the plant shoot under water
B ensuring that the entire system is kept airtight
C using a plant that bears both leaves and flowers
D preventing the air bubble from reaching and entering the shoot

12 The bubble in a bubble potometer moves most rapidly when the apparatus is in conditions which are
A dark and still. **B** dark and windy.
C light and still. **D** light and windy.

13 A weight atmometer loses LEAST weight when exposed to conditions which are
A cool and damp. **B** cool and dry.
C hot and damp. **D** hot and dry.

14 In which of the following do BOTH factors affect the rate of water uptake in a potometer but NOT in an atmometer?
A relative humidity and air movement
B light intensity and stomatal closure
C relative humidity and light intensity
D air movement and stomatal closure

15 The following graph charts the rate of transpiration from a geranium plant's leaves. When did the plant's stomata begin to open?

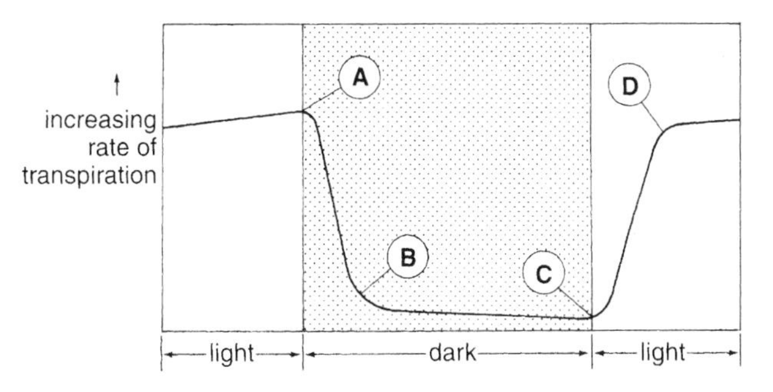

16 The table below gives the rates of transpiration for three different plants.

	rate of transpiration (mg water/cm² leaf/hr)	
	light	*dark*
holly	20	0
heather	10	0
sycamore	30	1

Which of the following is NOT a reasonable explanation of these results?

A some stomata remain partly open in darkness.
B cuticular transpiration occurs from some leaves.
C transpiration always comes to a halt in darkness.
D a little water escapes from a leaf's thin edges.

Items **17** and **18** refer to the following diagram of part of a leaf from a xerophyte.

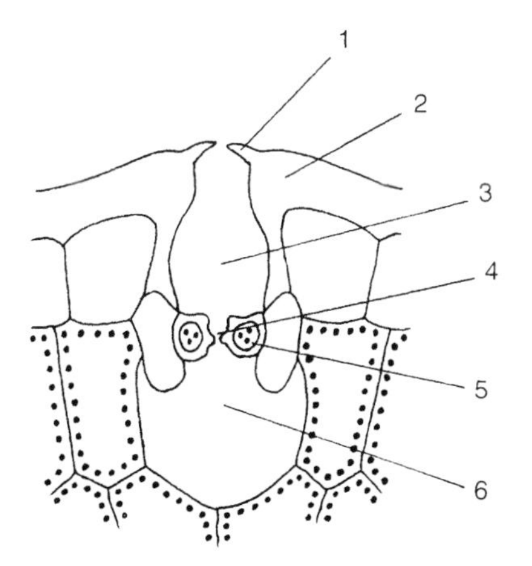

17 The sunken stoma is labelled
A 3. **B** 4. **C** 5. **D** 6.

18 Which of the following is NOT a xerophytic feature?
A 1 **B** 2 **C** 3 **D** 6

Items 19 and 20 refer to the following diagram of the pondweed *Potamogeton*.

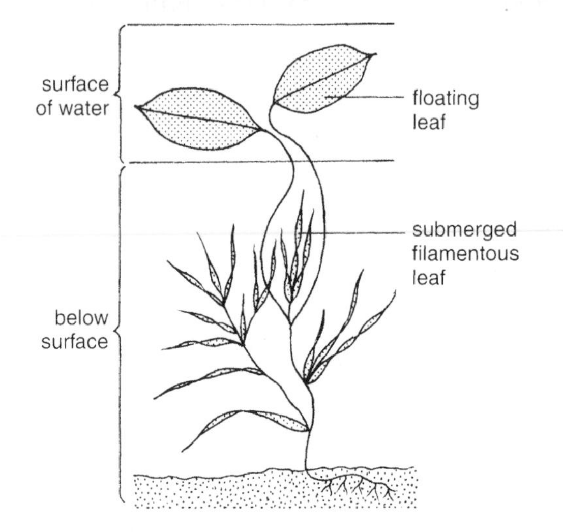

19 Which of the descriptions in the following table refers to a floating leaf?

	number of stomata on upper surface	*number of stomata on lower surface*
A	many	none
B	none	many
C	none	none
D	many	many

20 Which of the following statements is FALSE?

The submerged filamentous leaves
A offer little resistance to rapid water flow.
B present a large surface area for light absorption.
C possess many xylem vessels for support.
D take in mineral salts over a big surface area.

24 Obtaining food – animals

Match the terms in list X with their descriptions in list Y.

list X		list Y	
1	competitive exclusion	**a**	social signal used by a low-ranking member of a social hierarchy to indicate acceptance of the dominant leader
2	co-operative hunting	**b**	struggle for existence between members of different species when a resource is scarce
3	dominance hierarchy	**c**	area inhabited and defended by an individual animal or a breeding pair
4	interspecific competition	**d**	time spent by a foraging animal to locate food
5	intraspecific competition	**e**	result of interspecific competition where one species ousts another from an ecosystem
6	net energy gain	**f**	system of social organisation where the members are graded into a rank order
7	pursuit time	**g**	social signal used by the leader in a dominance hierarchy to assert authority
8	ritualised threat gesture	**h**	competition amongst members of the same species for territories
9	search time	**i**	type of foraging behaviour employed by a group of predators resulting in mutual benefits
10	subordinate response	**j**	time spent by a foraging animal to obtain food
11	territoriality	**k**	struggle for existence between members of the same species when a resource is scarce
12	territory	**l**	overall profit made by an animal when foraging costs are subtracted from benefits

Choose the ONE correct answer to each of the following multiple choice questions.

1 Which of the following diagrams BEST represents the trail followed by an ant when foraging for food and then, on finding it, returning to the colony?

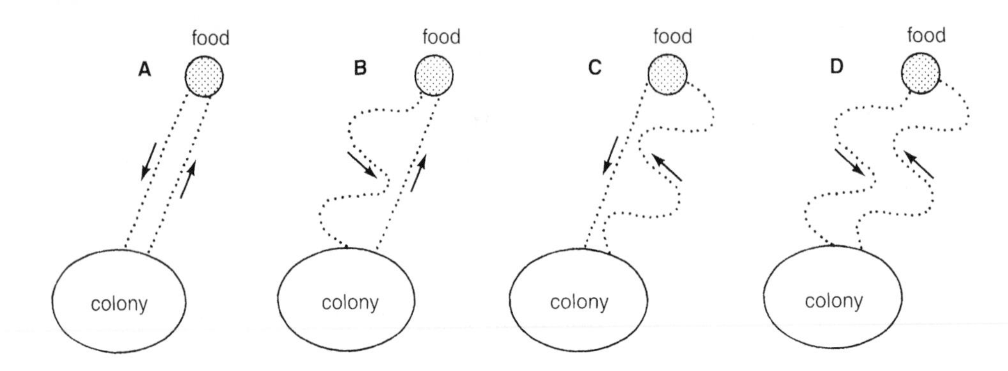

Items **2** to **4** refer to the following information and diagram.

On returning to the hive from a successful foraging trip, a bee does a waggle dance which communicates vital information to other bees as follows:

feature of dance	*information conveyed*
number of turns/min	distance of food from hive (the faster the dance, the closer the food)
indication of angle between sun and food supply	direction of food from hive
duration of dance	richness of food supply (the longer the dance, the richer the food)

The diagram on the next page shows a hive and six food supplies.

2 Which of the following is the richest source of food located furthest from the hive?
A P **B** Q **C** R **D** U

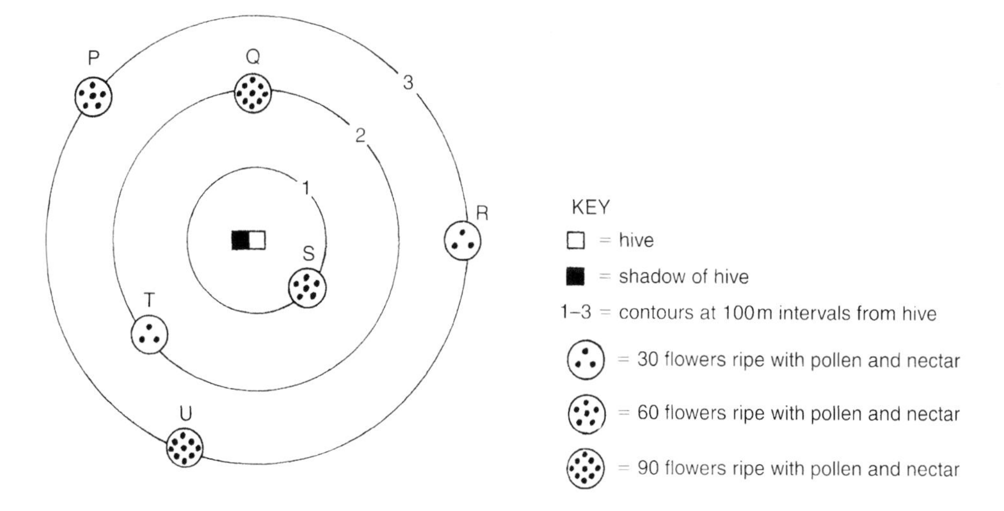

3 Which of the following angles (measured in a clockwise direction) would direct bees to food source P?

A 50° **B** 90° **C** 140° **D** 220°

4 Which of the following bees is indicating a food source NOT shown in the diagram?

bee	*speed of dance*	*angle (measured in clockwise direction)*	*duration of dance*
A	fast	240°	long
B	medium	140°	short
C	fast	40°	medium
D	slow	0°	short

5 Which of the following tabulated descriptions of foraging behaviour BEST applies to a pride of lionesses hunting in a savannah ecosystem rich in game?

	search time	*pursuit time*	*choice of prey*
A	short	long	selective
B	short	long	unselective
C	long	short	selective
D	long	short	unselective

6 Which of the following is NOT a feature of foraging behaviour?

A Energy can be saved by searching for food in a group.
B Energy supply can be maintained by defending a territory.
C Energy derived from prey must be equal to that expended in the search.
D Energy sources of low quality are often ignored in favour of high quality prey.

7 The accompanying graph shows the results from an investigation into the relationship between prey length and energy gained by a certain species of predatory bird.

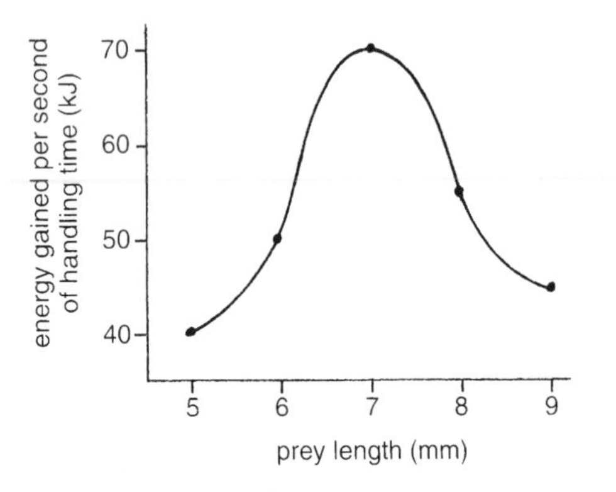

These results illustrate the principle that

A the smallest prey is the most profitable in terms of net energy gain since it is the easiest to catch and kill.
B the medium-sized prey provides the largest net energy gain since it contains most energy relative to the energy expended on the hunt.
C the largest prey contains most energy and offers the predator the greatest net energy gain for the effort required to subdue it.
D all sizes of prey provide similar net energy gains if an appropriate number are caught and consumed by the predator.

8 Competition between members of different species is called

A interspecific and is normally less intense than intraspecific competition.
B interspecific and is normally more intense than intraspecific competition.
C intraspecific and is normally less intense than interspecific competition.
D intraspecific and is normally more intense than interspecific competition.

9 Which of the following graphs illustrates the result of competition between a successful species and an unsuccessful species of animal for the same source of food?

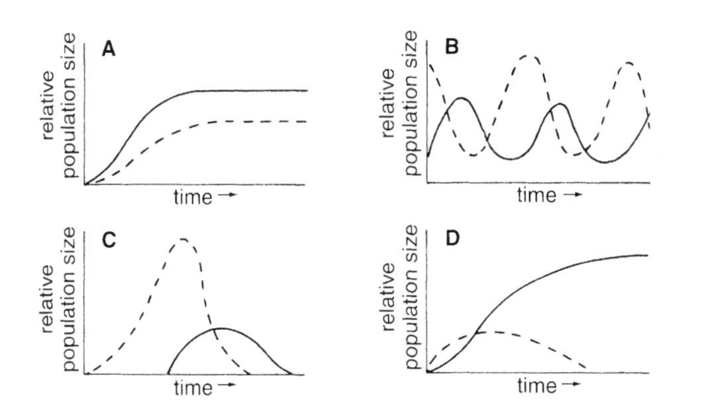

10 The diagrams below show the same area of moorland containing territories inhabited by red grouse in different years. Which diagram BEST represents the year in which the food supply was most plentiful?

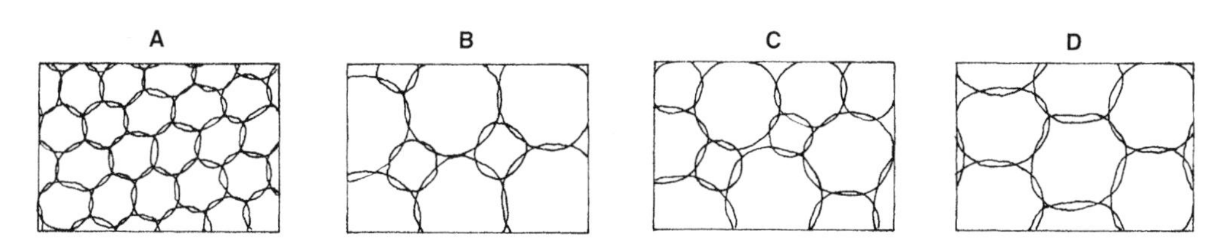

11 The graph below shows the increase in a population of fieldmice with time.

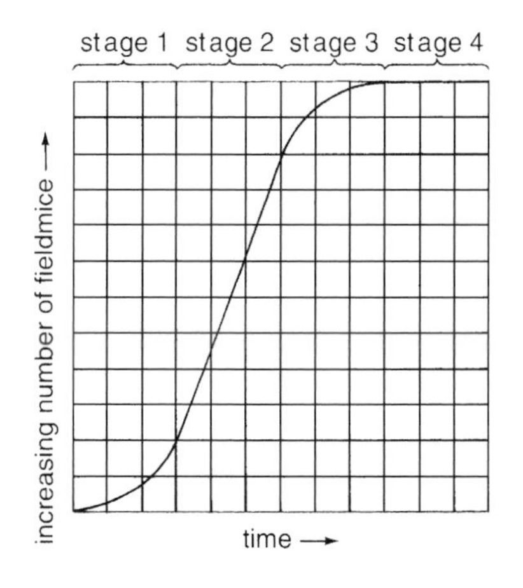

Which of the following explanations FAILS to account for the change occurring at the particular stage of the graph?

	stage	*possible explanation*
A	1	predators have been removed
B	2	food supply has become limiting
C	3	interspecific competition has increased
D	4	overcrowding has caused disease to spread

12 The following graph refers to the economics relating to the defence of different-sized territories by a certain species of animal.

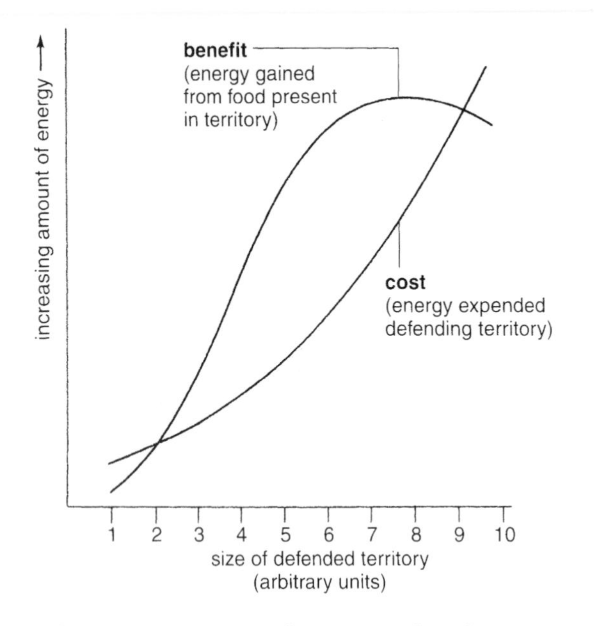

The optimum size of territory for this species under these conditions of food availability is

A 4. **B** 5. **C** 6. **D** 7.

13 The following list refers to pecking behaviour observed amongst six hens (P, Q, R, S, T and U).

P pecked U
R pecked T
S pecked U
U pecked R
P pecked T
S pecked P
T pecked Q
U pecked Q

Which bird was THIRD in the peck order?

A P **B** U **C** R **D** T

14 Which of the following does NOT occur as a direct result of a herd of animals being dominated by one older male?

A minimum aggression amongst the group members
B experienced leadership during times of crisis
C promotion of the group's chance of survival
D equal choice of food for all herd members

15 Which of the following are BOTH subordinate responses shown by a young wolf in the presence of the pack leader?

A head lowered and ears cocked
B hackles raised and tail lowered
C ears flattened and eyes averted
D eyes staring and teeth bared

25 Obtaining food – plants

Match the terms in list X with their descriptions in list Y.

list X		**list Y**	
1	autotrophic	**a**	arrangement of leaves which results in minimal overlap and maximum absorption of light
2	compensation point	**b**	autotroph that thrives in brightly illuminated habitats
3	heterotrophic	**c**	ability of a living organism to move its whole body from place to place
4	interspecific competition	**d**	the light intensity at which a plant's rate of photosynthesis equals its rate of respiration
5	intraspecific competition	**e**	autotroph that thrives in dimly lit habitats
6	mobility	**f**	mode of life where an organism remains fixed to one position in its habitat
7	mosaic pattern	**g**	term describing an organism that requires a ready-made source of food
8	sessility	**h**	struggle for existence between members of the same species when a resource is scarce
9	shade plant	**i**	term describing an organism able to make its own food by photosynthesis
10	sun plant	**j**	struggle for existence between members of different species when a resource is scarce

Choose the ONE correct answer to each of the following multiple choice questions.

Items **1** and **2** refer to the following possible answers.

A sessile and autotrophic
B sessile and heterotrophic
C mobile and autotrophic
D mobile and heterotrophic

1 Which description applies to the vast majority of flowering plants?

2 Which description is true of all advanced animals?

3 The leaves of many sun plants grow in such a way that each leaf blade is held at a certain angle to the rays of the midday sun.

Which of the following would be the most effective angle for light absorption?

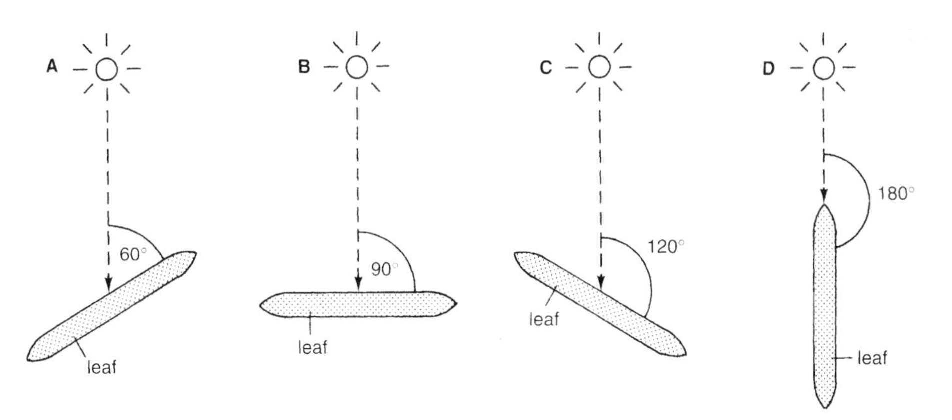

4 Competition between members of the same species is called

A intraspecific and is more intense than interspecific competition.
B intraspecific and is less intense than interspecific competition.
C interspecific and is more intense than intraspecific competition.
D interspecific and is less intense than intraspecific competition.

5 Which of the following leaf mosaic patterns would be the most effective?

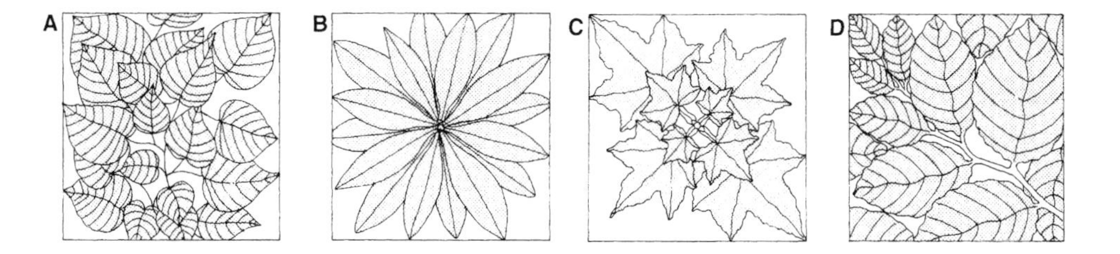

6 *Galium saxatile* grows best in acidic soil but will grow on alkaline soil in the absence of competition. *Galium pumilum,* on the other hand, grows best on alkaline soil but can survive on acidic soil.

In a competition experiment, equal numbers of seeds of both species of *Galium* were planted together in two pots. Pot 1 contained alkaline soil and pot 2 acidic soil.

Which of the following sets of results is the most likely outcome of this experiment?

	pot 1		*pot 2*	
	G. saxatile	*G. pumilum*	*G. saxatile*	*G. pumilum*
A	+	−	−	+
B	−	+	+	−
C	+	−	+	−
D	−	+	−	+

(+ = growth; − = no growth)

7 The graph below refers to the net amount of sugar produced or used in a green plant leaf over a 24-hour period.

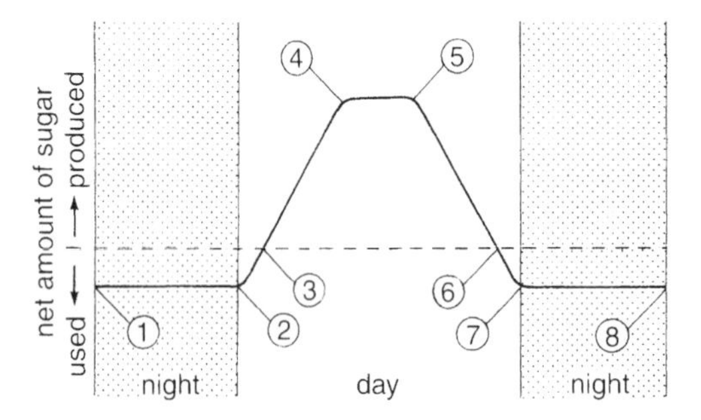

Which of the following are BOTH compensation points?

A 1 and 8 **B** 2 and 7 **C** 3 and 6 **D** 4 and 5

8 The following list gives four situations that could affect a mixed community of plants growing in a grassland ecosystem.

1 herbivores graze unselectively on all species
2 all herbivores are removed from the ecosystem
3 herbivores graze selectively on delicate plant varieties
4 herbivores graze selectively on sturdy dominant grasses

Which two situations would tend to maintain species diversity amongst the plant community?

A 1 and 3 **B** 2 and 3 **C** 1 and 4 **D** 2 and 4

9 With reference to the accompanying graph, it is CORRECT to say that

A X is the sun plant and its compensation point occurred at 3 units of light.
B X is the shade plant and its compensation point occurred at 5 units of light.
C Y is the sun plant and its compensation point occurred at 5 units of light.
D Y is the shade plant and its compensation point occurred at 3 units of light.

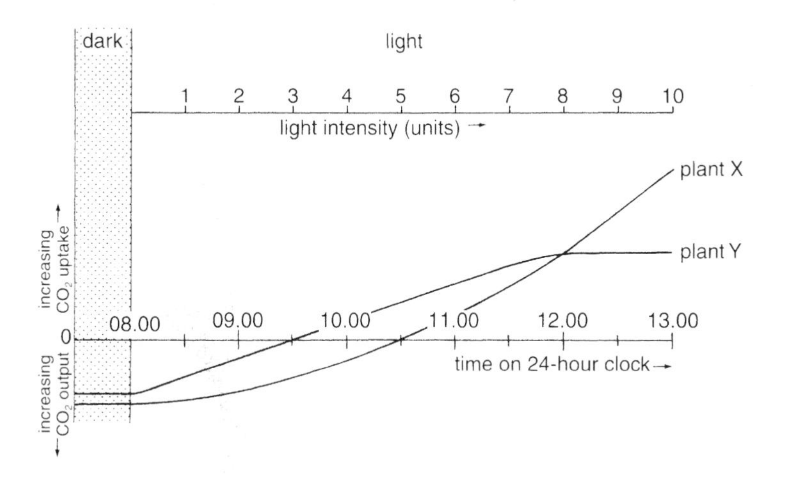

10 The leaf discs in the accompanying diagram have sunk to the bottom of the liquid in the syringe barrels because the air in their air sacs has been drawn out.

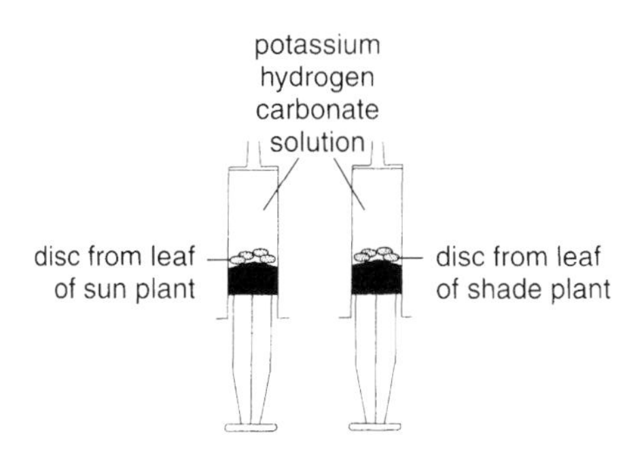

Which of the following diagrams best represents the location of the leaf discs after a period of illumination in green light?

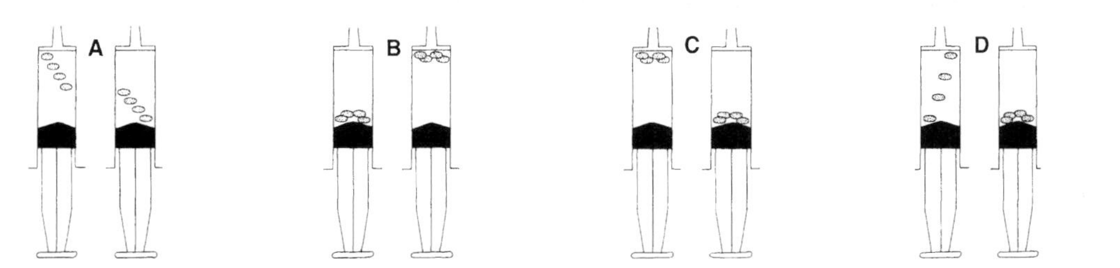

26

Coping with dangers

Match the terms in list X with their descriptions in list Y.

list X	list Y
1 alarm call	**a** process by which an animal learns not to respond to a harmless stimulus
2 avoidance behaviour	**b** social defence mechanism enabling musk oxen to fend off carnivorous predators
3 camouflage	**c** means by which an animal attempts to be ignored by a predator which only eats freshly killed prey
4 deflection display	**d** general term for behaviour involving modification of the response made to a stimulus
5 distraction display	**e** structural adaptations possessed by some insects which can be used to startle a predator
6 feigning death	**f** auditory warning signal used as a social defence mechanism by many birds and mammals
7 habituation	**g** use of bogus eye spots to deceive predators about the true identity of a prey's head and tail ends
8 individual defence mechanism	**h** means by which a parent bird diverts a predator's attention from the young in a nest on the ground
9 learning	**i** general term for the means by which an animal copes with danger as a member of a co-operative group
10 mobbing	**j** general term for a response made by an animal in order to escape potential danger
11 social defence mechanism	**k** passive defence mechanism which enables an animal to blend effectively with its background
12 wing markings	**l** general term for the means by which a solitary animal copes with danger

Choose the ONE correct answer to each of the following multiple choice questions.

Items **1** and **2** refer to the following information.

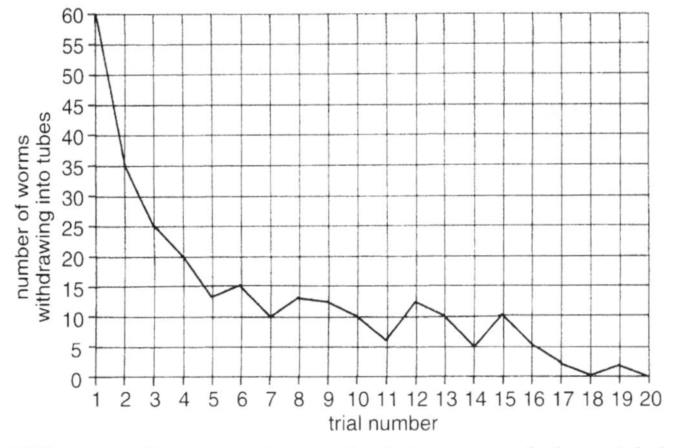

When a tube worm is touched, it responds by withdrawing into its tube. The graph above charts the results from an experiment where 60 tube worms were subjected to repeated touching over 20 trials.

1 At which trial number did 25% of the worms show the escape response?
A 3 **B** 6 **C** 10 **D** 15

2 At which trial number had 35 of the worms become habituated?
A 2 **B** 3 **C** 4 **D** 5

3 Habituation is a beneficial form of behaviour because it enables an animal to
A avoid a potentially dangerous situation.
B defend itself against an enemy.
C effect its escape response quickly.
D conserve energy for essential activities.

4 It is essential that habituation is a short-lived form of learning otherwise the animal could
A be left open to danger.
B expend energy needlessly.
C exhaust its food reserves.
D forget how to effect its escape response.

5 Which of the following would be most suitable for use as a finger maze to investigate the effect of experience on learning a new skill?

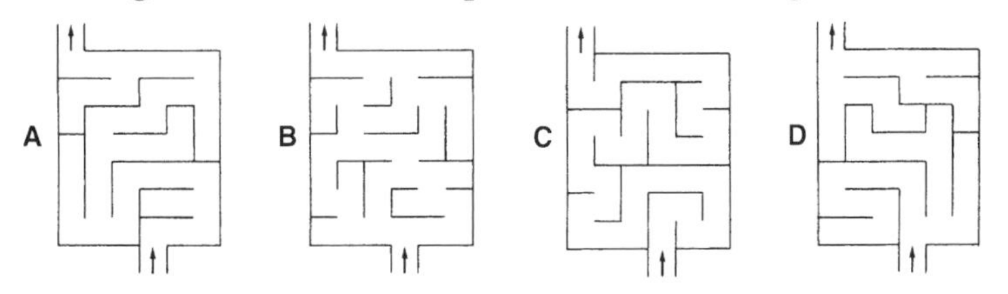

6 The following table lists some aspects of good practice carried out during a finger maze investigation. Which practice is correctly paired with the reason for carrying it out?

	good practice	*reason*
A	10 trials per learner	to ensure that no second variable factor is included in the investigation
B	experiment repeated with many learners	to obtain a more reliable set of results
C	learner blindfolded throughout all 10 trials	to prevent two fingers being used at once
D	same learner used for each group of 10 trials	to ensure that the recorded results are accurate

7 Which of the following graphs correctly represents a learning curve?

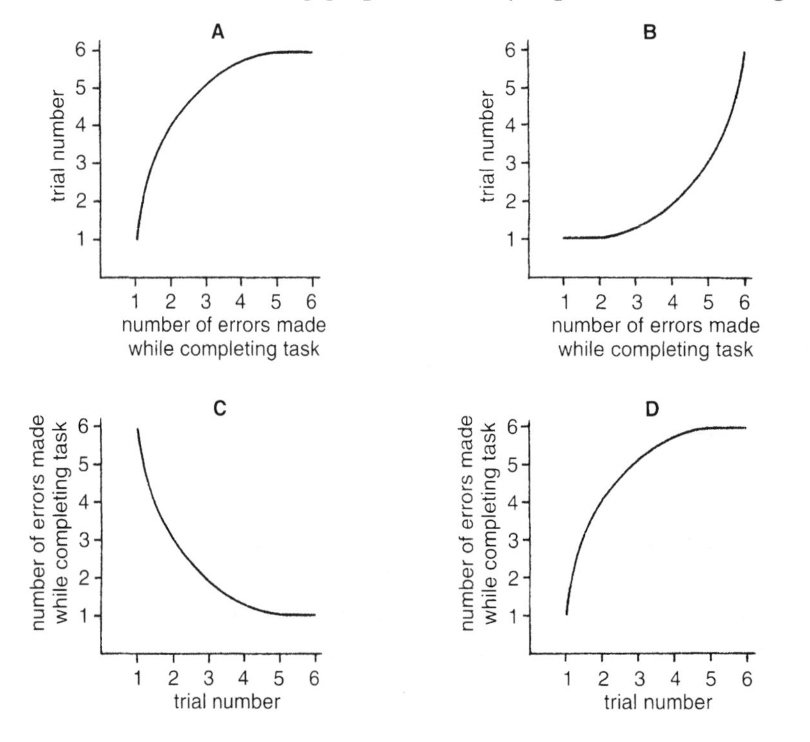

8 Young children are taught by their parents not to poke their fingers into electrical sockets. To be of protective value it is essential that such learning involves a
A short-term modification of the response made to the stimulus.
B long-term modification of the response made to the stimulus.
C short-term modification of the stimulus made to the response.
D long-term modification of the stimulus made to the response.

Items **9**, **10** and **11** refer to the following possible answers.
A performing a 'broken wing' distraction display
B rolling over and feigning death
C fleeing from the enemy at top speed
D possessing a body resembling nearby plants

9 Which of these is a passive form of defence in animals?

10 Which of these defence mechanisms is employed by a grass snake?

11 Which of these defence mechanisms is employed by a gazelle?

Items **12** and **13** refer to the following possible answers.
A mimicry
B deflection display
C counter shading
D menacing eye markings

12 Which of these is a defence mechanism possessed by a trout?

13 Which of these is a defence mechanism possessed by non-poisonous striped insects such as the robber fly?

14 Which of the following is a social mechanism for defence?
A alarm calls in birds
B spines on bodies of hedgehogs
C poison glands in snakes
D foul-smelling secretions squirted by skunks

15 Which of the following is NOT a social mechanism for defence?
A mobbing employed by musk oxen when under attack
B dominance hierarchy observed by baboons when on the move
C circular formation adopted by bobwhite quails when resting
D injection of poison into enemies by adders when threatened

16 The diagram illustrates a desert plant called *Echinocactus*.

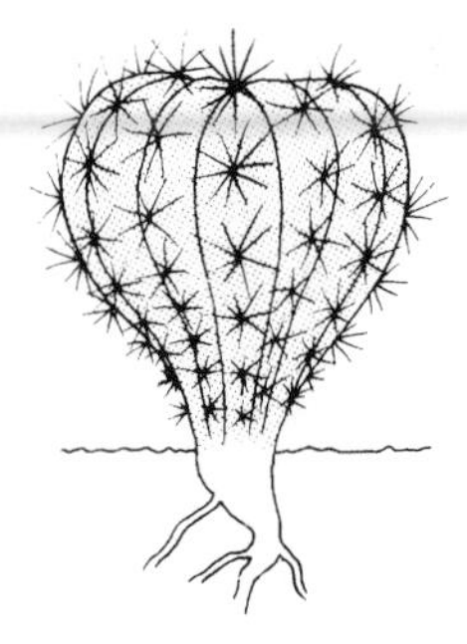

This cactus is protected from thirsty mammals by possessing
A a rounded shape.
B a thick waxy cuticle.
C leaves reduced to spines.
D water stored in succulent tissues.

Items **17** and **18** refer to the labelled structures in the following diagram of four different types of plant.

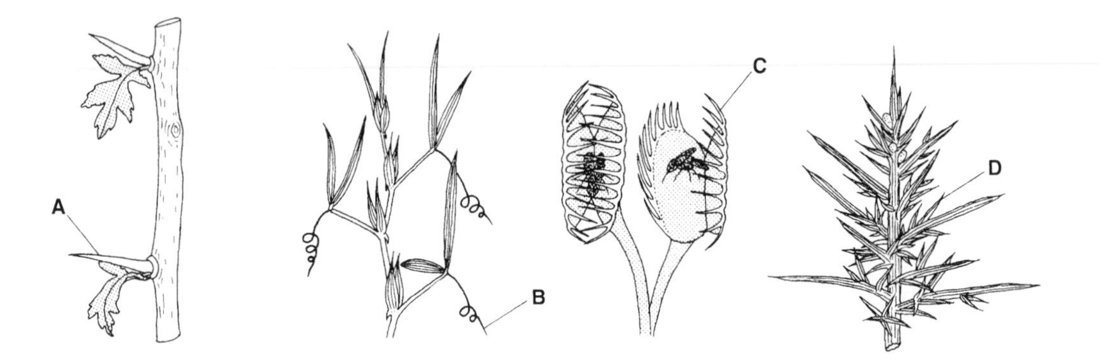

17 Which structure is a protective spine?

18 Which structure is a thorn?

19 The diagram below shows four common weeds (not drawn to scale). Which would be LEAST likely to survive on a meadow heavily grazed by cattle?

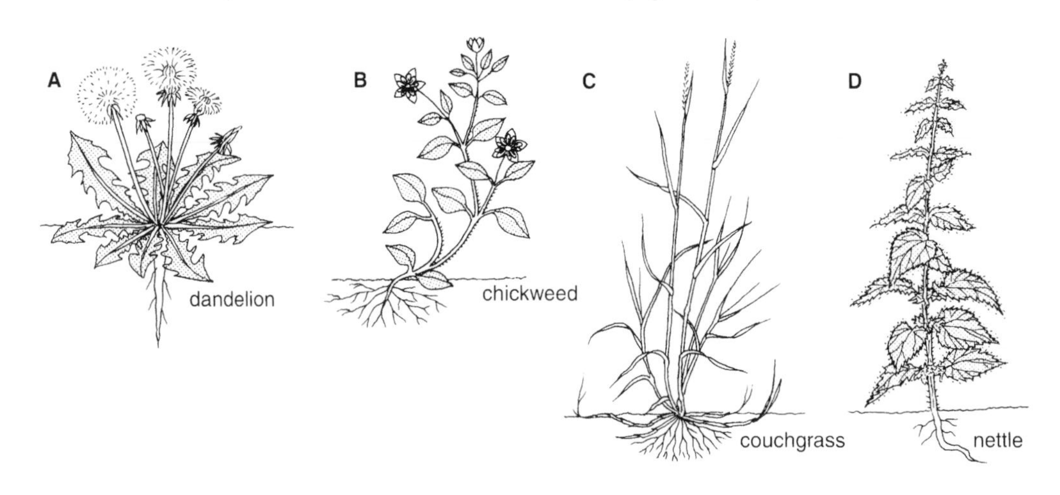

20 Which of the following graphs BEST represents the relationship between number of spines per holly leaf and height of leaf from ground?

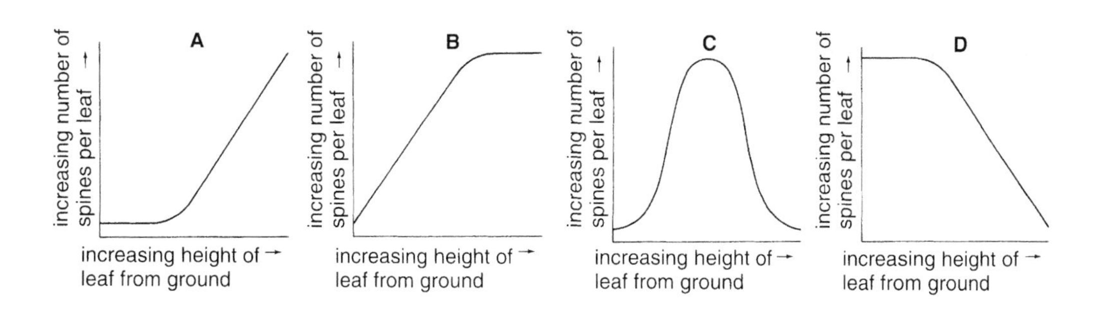

27

Plant growth

Match the terms in list X with their descriptions in list Y.

list X	**list Y**
1 apical meristem	**a** process by which an organism replaces lost or damaged parts
2 autumn wood	**b** process by which newly formed cells at a meristem increase in length
3 cambium	**c** region of an annual ring of thickening containing large xylem vessels
4 differentiation	**d** form of artificial propagation used to produce a clone of a desirable plant
5 elongation	**e** general name for a group of undifferentiated plant cells capable of dividing repeatedly
6 growth	**f** region of an annual ring of thickening containing small xylem vessels
7 lateral meristem	**g** site of primary growth in a plant such as a root tip or shoot tip
8 meristem	**h** process by which an unspecialised cell becomes structurally adapted to perform a special function
9 regeneration	**i** type of lateral meristem responsible for secondary thickening of a woody plant's stem
10 spring wood	**j** process by which newly formed cells at a meristem develop a permanent central vacuole
11 tissue culture	**k** site of secondary growth in a plant
12 vacuolation	**l** irreversible increase in dry mass of an organism

Choose the ONE correct answer to each of the following multiple choice questions.

1 Meristematic cells are
A found only at root and shoot tips in plants.
B undifferentiated and capable of dividing repeatedly.
C widely distributed throughout a developing animal's body.
D used to transport materials in a plant's stem.

Items 2 and 3 refer to the following information.

During growth at a plant apex, each cell undergoes the following processes.
1 differentiation **2** elongation
3 division **4** vacuolation

2 The order in which these occur is
A 3, 2, 4, 1. **B** 2, 3, 4, 1.
C 3, 2, 1, 4. **D** 2, 3, 1, 4.

3 Increase in the length of a plant depends on the occurrence of BOTH
A 1 and 2. **B** 1 and 3.
C 2 and 3. **D** 1 and 4.

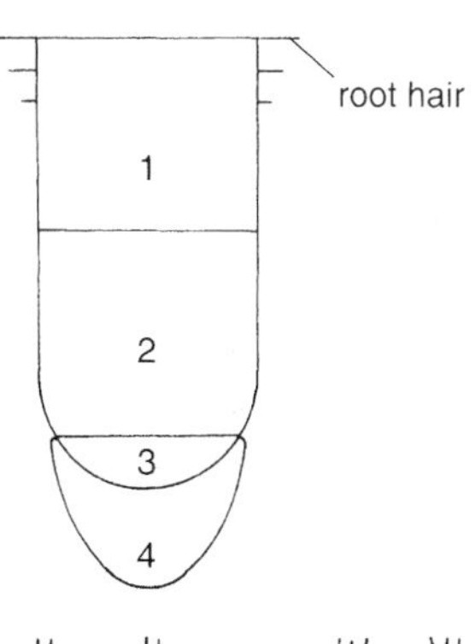

Items 4 and 5 refer to the diagram of a root tip opposite.

4 The diagram below it shows a certain type of cell undergoing change during its development.

In which region of the root would this process of change occur?
A 1 **B** 2 **C** 3 **D** 4

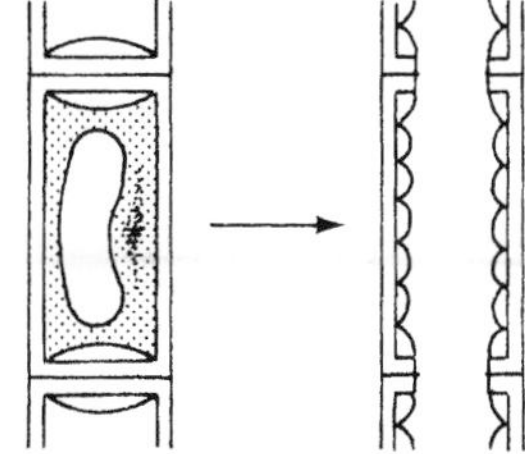

5 The accompanying diagram shows the appearance of two cells (X and Y) at different stages of their development.

In which regions of the root would such cells be found?

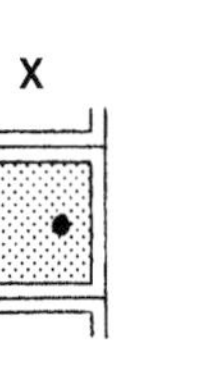

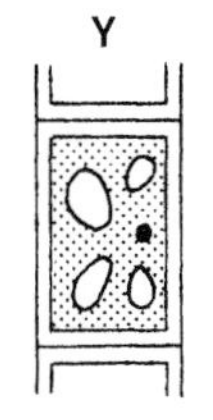

	X	Y
A	2	1
B	2	3
C	3	2
D	4	3

6 The diagram opposite shows a transverse section through a shoot apex. Which structure gives rise to leaf primordia?

7 The diagram opposite shows a longitudinal section through a shoot apex.

leaf primordium
young leaf
lateral bud

Which of the diagrams below shows the correct appearance of this shoot apex at the formation of the next leaf primordium?

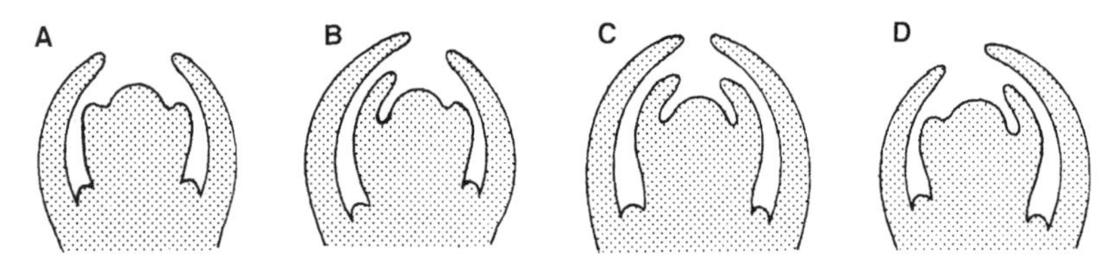

8 The diagram below (left) shows the shoot apex of a young plant marked at regular intervals with waterproof ink. Which of the diagrams below (A, B, C or D) best represents the marked region of this shoot apex after several days of growth?

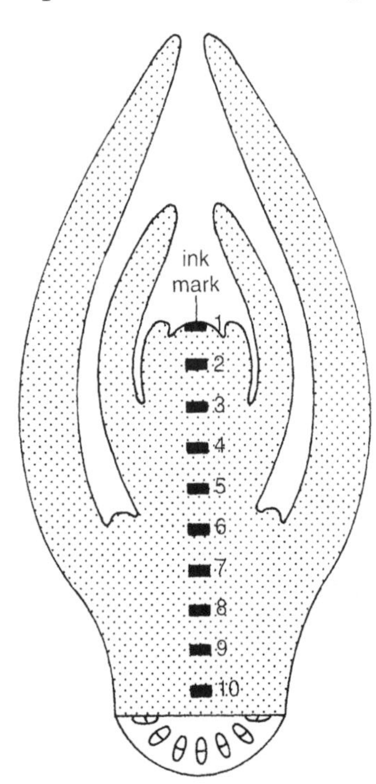

Items **9**, **10** and **11** refer to the accompanying diagram which shows a transverse section of a woody stem.

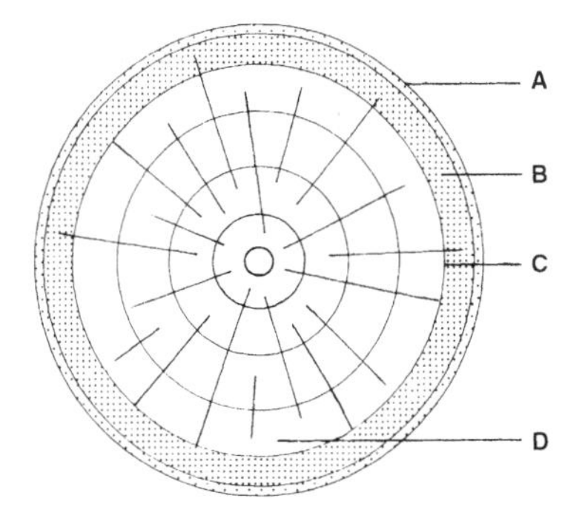

9 Which letter indicates xylem vessels?

10 Which letter indicates phloem?

11 Which letter indicates cambium?

12 A piece of wood, cut from a felled tree, is shown in the diagram.

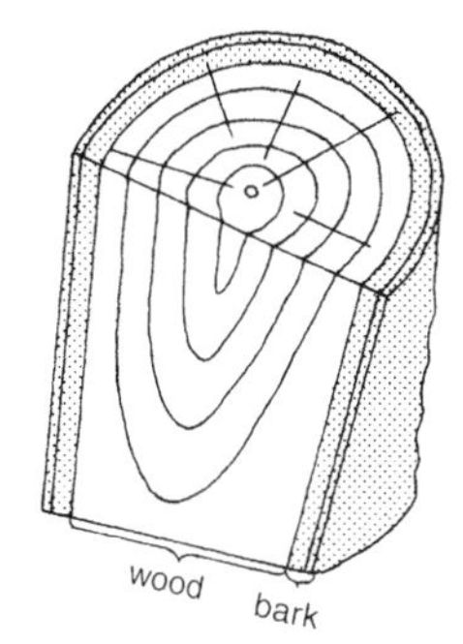

The age of the tree in years is

A 4 **B** 5 **C** 6 **D** 7

13 From which region of the following woody stem has sample X been taken?

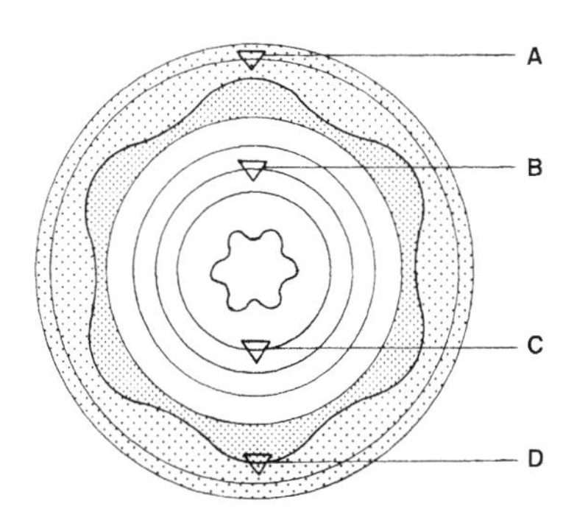

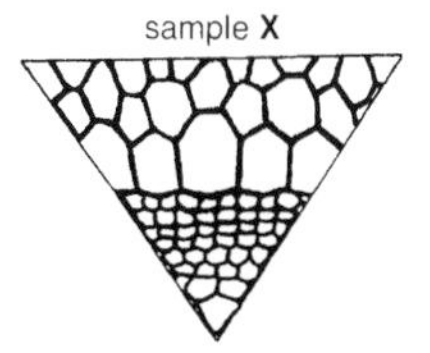

14 Which combination of conditions during a spring and summer would result in the development of the narrowest ring of secondary thickening in a woody stem?

A warm weather, infestation by greenfly and lack of rainfall
B cold weather, infestation by greenfly and abundant sunshine
C warm weather, lack of rainfall and abundant sunshine
D cold weather, lack of rainfall and infestation by greenfly

15 A forest ecosystem's annual rainfall figures for four years are shown in the following table.

year	*annual rainfall (mm)*
1983	813
1984	978
1985	826
1986	672

A tree growing in this region was cut down in December 1986. Which of the following diagrams best represents a transverse section of this tree?

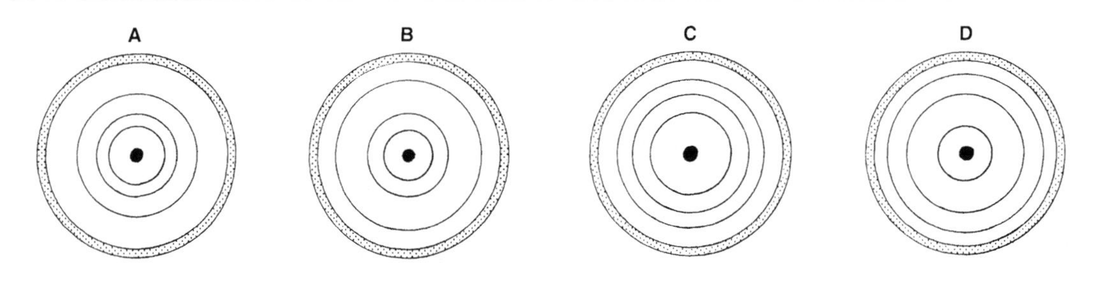

16 Which longitudinal cut through the tree shown (in transverse section) in the following diagram would produce the grain pattern?

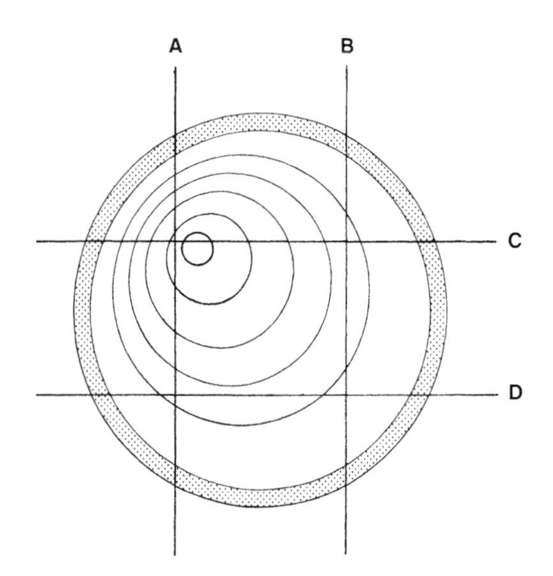

grain pattern

17 All of the transverse sections shown in the following diagram are from the one plant.

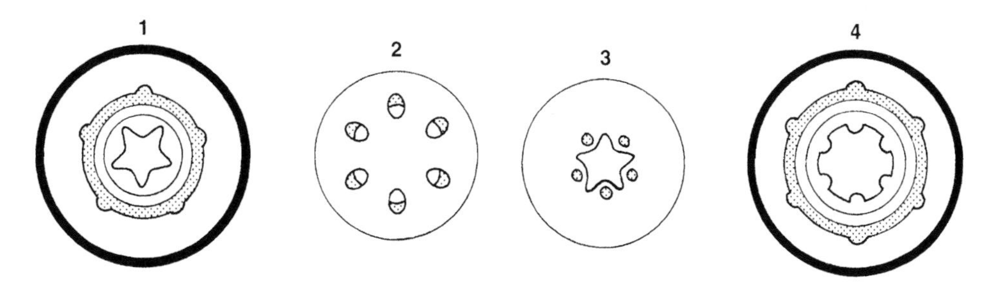

Water entering the plant at a root hair would pass through these in the order
A 1, 3, 4, 2. **B** 1, 4, 3, 2. **C** 3, 1, 4, 2. **D** 3, 2, 1, 4.

18 A clone of plants can be produced by EITHER
A taking cuttings OR growing tissue cultures.
B growing tissue cultures OR crossing two different varieties.
C crossing two different varieties OR applying growth hormones to seedlings.
D applying growth hormones to seedlings OR taking cuttings.

19 Which of the following tissues CAN be regenerated by the human body by natural means?
A muscle **B** heart **C** bone **D** brain

20 Which of the following cell types CANNOT be regenerated by the human body by natural means?
A blood **B** lung **C** liver **D** skin

28 Growth patterns

Match the terms in list X with their descriptions in list Y.

list X	list Y
1 adolescent growth spurt	**a** describing a discontinuous growth pattern
2 continuous	**b** rapid phase of human growth during early childhood
3 dry mass	**c** S-shaped graph showing an organism's four periods of growth
4 ecdysis	**d** unreliable indicator of growth
5 exoskeleton	**e** describing a smooth growth pattern
6 fresh mass	**f** rapid phase of human growth during puberty
7 infant growth spurt	**g** moulting of skin by an insect to allow its body length to increase
8 intermittent	**h** reliable indicator of growth
9 sigmoid growth curve	**i** hard inelastic skin preventing continuous increase in length of an insect's body

Choose the ONE correct answer to each of the following multiple choice questions.

1 Which of the following is the most reliable indicator of growth of an annual plant (e.g. broad bean)?

A fresh mass **B** dry mass
C shoot length **D** root length

Items 2 to 4 refer to the following diagram which shows a typical S-shaped growth curve.

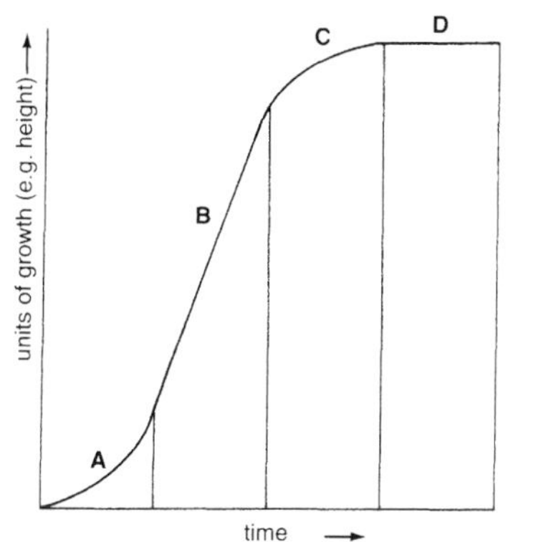

2 Which region of the graph represents the period of decelerating growth?

3 Which region of the graph represents the period of accelerating growth?

4 Which region of the graph represents the period of no growth?

Items 5 to 7 refer to the following four graphs.

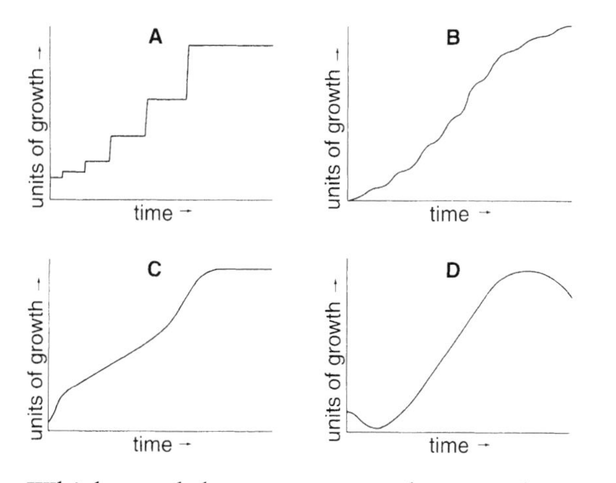

5 Which graph best represents the growth curve of an oak tree's height?

6 Which graph best represents the growth curve of a cockroach's body length?

7 Which graph best represents the growth curve of a human being's body mass?

8 The following table compares the growth of a rose bush with that of a gerbil. Which pair of statements is INACCURATE?

	gerbil	*rose bush*
A	increase in size stops on reaching adulthood	increase in size continues throughout life
B	growth occurs all over body	growth occurs only at meristems
C	does not show an increase in dry mass	does show an increase in dry mass
D	regenerative powers are very limited	regenerative powers are fairly extensive

Items **9** and **10** refer to the accompanying graph which shows the rate of increase in height of boys and girls between the ages of six months and 18 years (based on data from a large population).

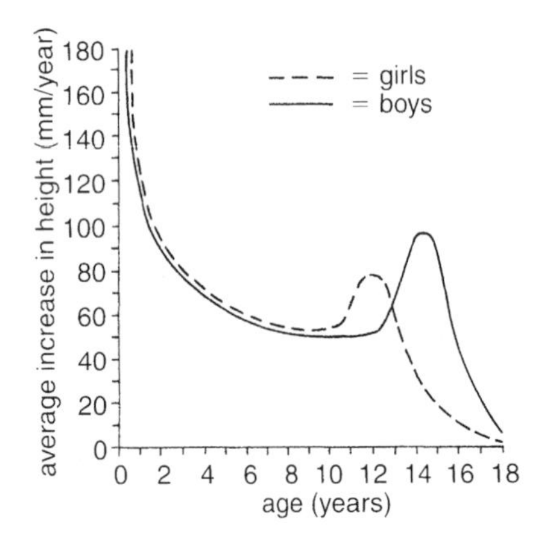

9 Between which ages did the boys gain height at the fastest rate?

A 10–11
B 11–12
C 12–13
D 13–14

10 On average the girls showed an annual gain in height of 80 mm at ages

A 3 and 12.
B 3 and 13.
C 3 and 15.
D 13 and 15.

11 The following graph charts the growth in length of a human foetus before birth.

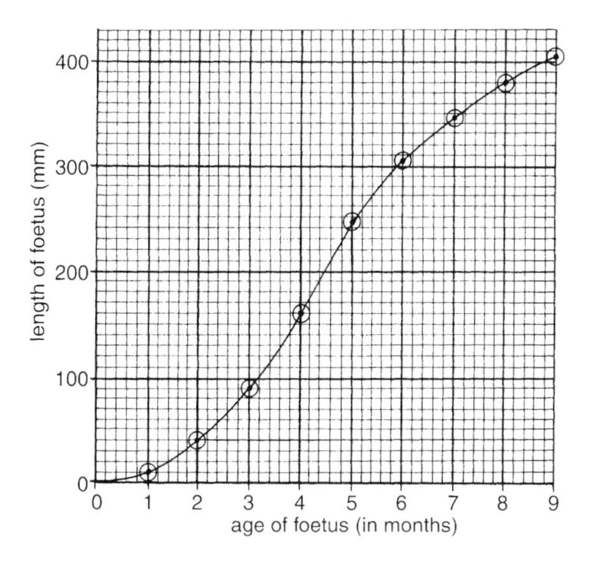

What was the average rate of growth of the foetus in mm/month during the final four months of the pregnancy?

A 25 **B** 40 **C** 160 **D** 405

12 The data in the table below refer to a small mammal.

age (days)	*mass (g)*	*gain in mass since previous weighing (g)*	*length of time interval between weighings (days)*	*average daily gain in mass (g)*
3	3	–	–	–
6	4	1	3	0.33
12	10	6	6	1.00
28	30	20	16	box Y
38	54	box X	10	2.40

The figures in boxes X and Y should be

	X	Y
A	27	1.25
B	27	1.33
C	24	1.25
D	24	1.33

13 In an insect's life cycle, each stage which occurs between two moults (ecdyses) is called an instar. A locust passes through several instar stages before reaching its final size.

The graph below shows the effect of temperature on rate of development in the locust.

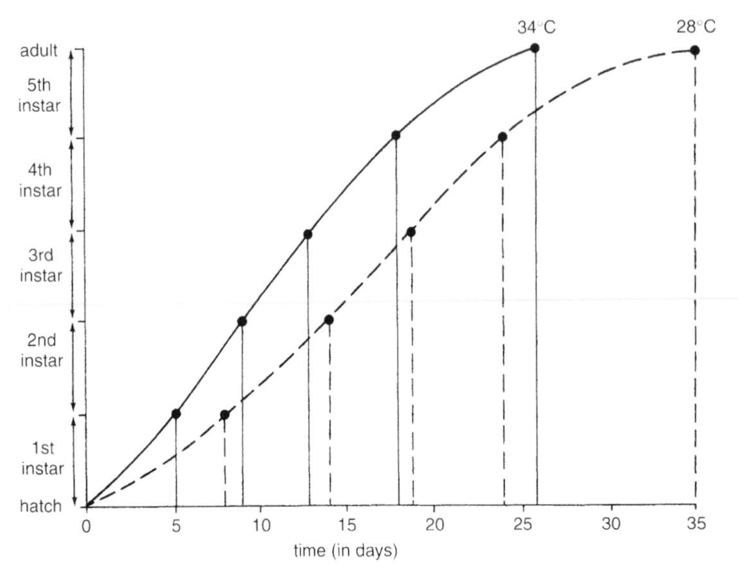

Which of the following is NOT true?

A Insects at the lower temperature take nine days longer to reach the adult stage.
B Insects at the higher temperature have reached the fifth instar after 18 days.
C At the lower temperature insects take three days longer to reach the second instar.
D At both temperatures insects have reached the fourth instar after 15 days.

Items **14** and **15** refer to the table below which gives the average height of human males at different ages.

age (in years)	*height (in mm)*
birth	506
2	875
4	1034
6	1175
8	1300
10	1403
12	1496
14	1627
16	1716
18	1745

14 Maximum increase in height occurs between
A birth and 2 years. **B** 4 and 6 years.
C 14 and 16 years. **D** 16 and 18 years.

15 A histogram drawn from these data by plotting increase in height against age would have the appearance

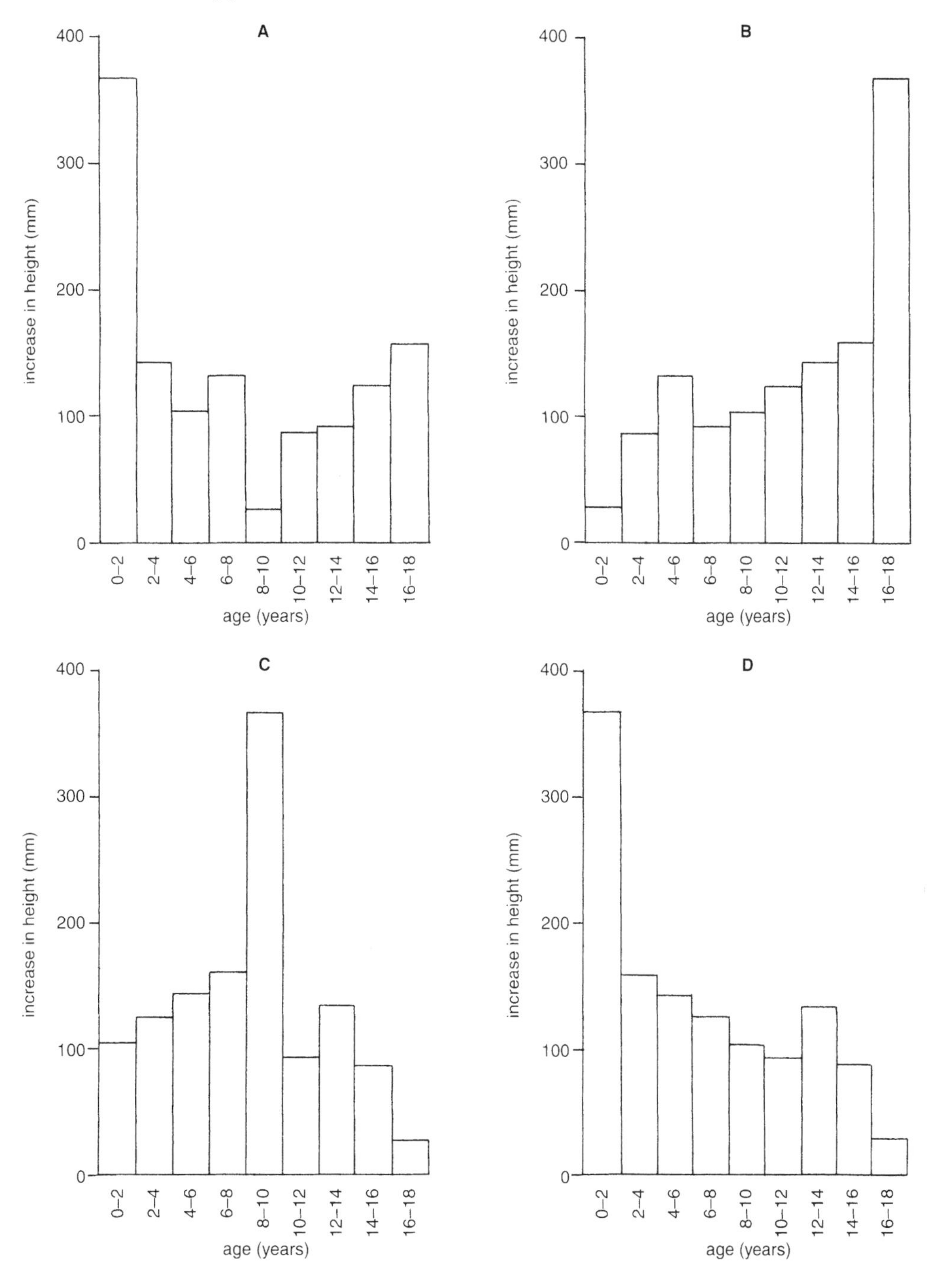

29 Genetic control

Match the terms in list X with their description in list Y.

list X		list Y	
1	albinism	**a**	region of DNA which codes for a repressor molecule
2	differentiation	**b**	molecule coded for by a regulator gene which can bind with an operator gene
3	enzyme	**c**	region of DNA which codes for a functional protein (e.g. an enzyme)
4	inducer	**d**	region of DNA which acts as a site for attachment of repressor molecule
5	metabolism	**e**	molecule which prevents a repressor molecule from blocking an operon
6	metabolite	**f**	protein molecule which acts as a biological catalyst
7	mutation	**g**	genetically inherited condition in which the sufferer lacks the enzyme needed to make melanin
8	operator	**h**	a change in an organism's genetic material
9	phenylketonuria	**i**	process of cell specialisation involving the selective switching off or on of certain genes
10	regulator gene	**j**	sum of all the chemical reactions occurring within a living organism
11	repressor	**k**	substance that takes part in a metabolic process by being the product of one enzyme and the substrate of another
12	structural gene	**l**	genetically inherited condition in which the sufferer lacks the enzyme needed to convert phenylalanine to tyrosine

Choose the ONE correct answer to each of the following multiple choice questions.

1 The chemical reaction catalysed by the enzyme β-galactosidase is

A lactose → glucose + galactose **B** galactose → lactose + glucose

C lactose → glucose + maltose **D** galactose → lactose + maltose

Items 2, 3 and 4 refer to the diagram below which shows a possible arrangement of the genes involved in the induction of the enzyme β-galactosidase in *Escherichia coli.*

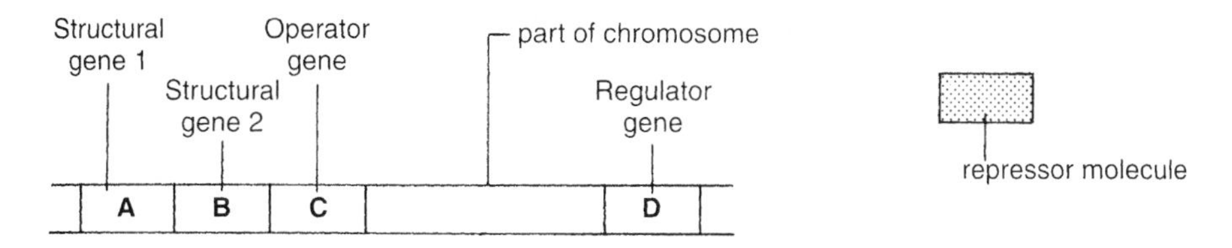

2 To which gene would the repressor molecule become attached in the absence of lactose?

3 Which gene produced the repressor molecule?

4 In this system the inducer molecule is called

A lactose. **B** galactose. **C** glucose. **D** β-galactosidase.

Items 5, 6 and 7 refer to the following information, diagram and table of possible answers.

Tryptophan is an amino acid needed for the synthesis of proteins. Situation 1 in the diagram shows a set of circumstances where a structural gene remains switched on in a cell of *Escherichia coli.*

Under different circumstances, the series of events shown in situation 2 is thought to occur. This brings about the repression of the synthesis of an enzyme.

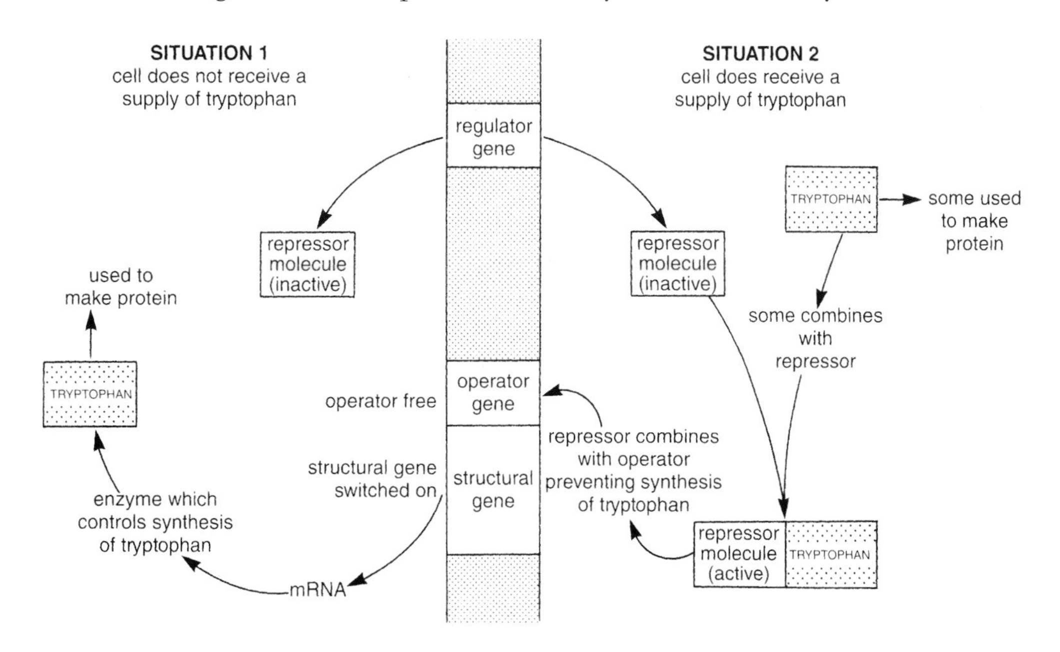

	Regulator gene	*Operator gene*	*Structural gene*
A	on	blocked	off
B	off	blocked	off
C	off	free	on
D	on	free	on

5 Which answer refers to the state of the genes in a cell of *Escherichia coli* grown on nutrient agar lacking tryptophan?

6 Which answer refers to the state of the genes in a cell of *Escherichia coli* cultured in nutrient broth containing tryptophan?

7 Which answer refers to the situation that would arise if the supply of tryptophan referred to in Item 6 ran out?

8 The following statements describe possible benefits to a cell from being able to switch genes on and off as required. Which statement is INCORRECT?
A A steady state is maintained despite fluctuations in supply of materials.
B A constant supply of energy is made available for diffusion of ions to occur.
C Needless wastage of resources such as amino acids is prevented.
D Efficient use is made of available energy during cell metabolism.

9 The diagram below shows a simplified version of a biochemical pathway that occurs during cell metabolism in normal humans.

Which gene has undergone a mutation in a sufferer of phenylketonuria?

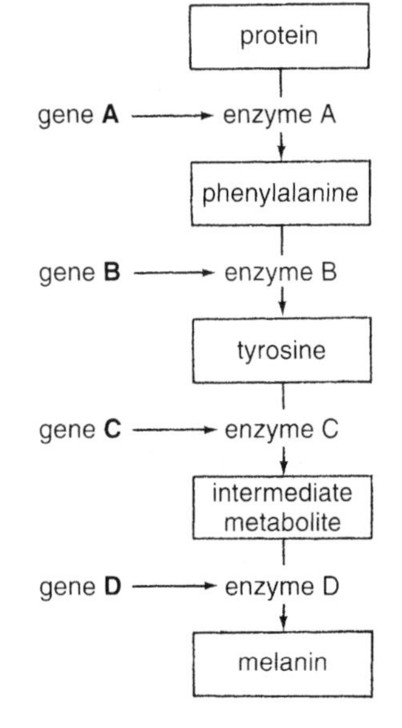

Items **10, 11** and **12** refer to the following diagram which represents the last four stages in a metabolic pathway in the fungus *Neurospora.*

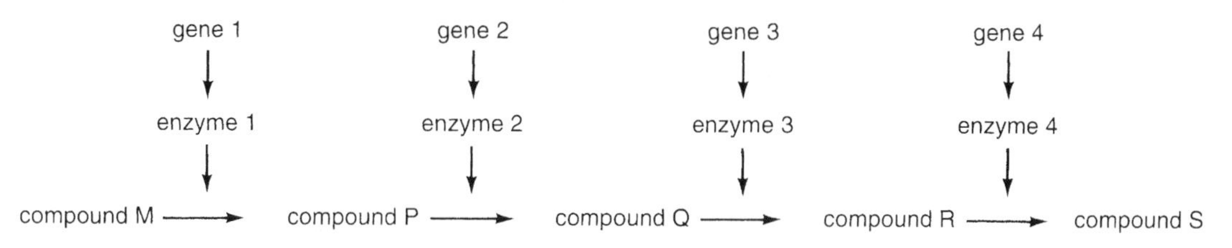

10 A mutant strain of the fungus is found to accumulate compound Q as a result of its metabolism. The gene which has undergone a mutation in this strain is
A 1 **B** 2 **C** 3 **D** 4

11 Wild type *Neurospora* can grow on minimal medium (sucrose, mineral salts and one vitamin) but mutant strains suffering metabolic blocks are unable to do so.

In an experiment the mutant strain referred to in Item 10 was subcultured onto the following plates.

plate
P = minimal medium + substance P
Q = minimal medium + substance Q
R = minimal medium + substance R
S = minimal medium + substance S

It would grow successfully on BOTH plates
A P and Q. **B** Q and R. **C** R and S. **D** S and P.

12 A different mutant strain was found to grow successfully on plate S (minimal medium + substance S) but on no other.

The enzyme that this strain *of Neurospora* fails to make is
A 1. **B** 2. **C** 3. **D** 4.

13 Each cell in a multicellular organism contains all the genes necessary for the construction of
A that one cell only.
B all cells of that type of tissue.
C all types of that cell.
D the whole organism.

14 The human gastric glands contain pepsin-secreting cells. Within each of these differentiated cells

A all genes continue to operate.
B certain genes no longer operate.
C certain genes are no longer present.
D none of the genes operate.

15 The diagram below shows a simplified version of the means by which 'Dolly, the sheep' was produced.

Which of the following statements is FALSE?

A Dolly and her mother are genetically identical members of a clone.
B A mammary gland cell in a sheep contains repressed genes that can become switched on again.
C Differentiation of the mammary gland cells in a sheep is an irreversible process.
D During Dolly's embryonic development, cells became differentiated and specialised.

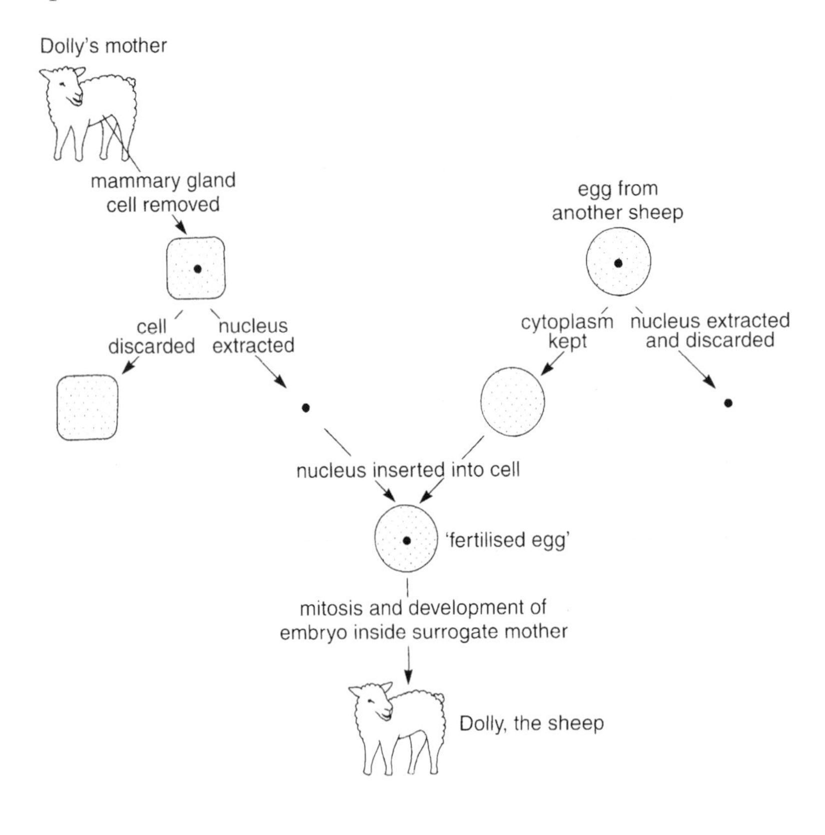

16 Which of the following structures BOTH contain pools of undifferentiated ('stem') cells which give rise to human blood cells?

A red bone marrow and spleen
B spleen and heart
C heart and lymph glands
D lymph glands and liver

17 In a phagocyte, some of the genes that are 'switched on' are those required for the formation of

A antibodies.
B digestive enzymes.
C haemoglobin.
D blood-clotting chemicals.

18 In a lymphocyte, some of the genes that are 'switched on' are those required for the formation of

A antibodies.
B digestive enzymes.
C haemoglobin.
D blood-clotting chemicals.

19 In a horse chestnut tree, the cells in a leaf of a newly opened bud do not give rise to a stem or root because such cells

A only possess the genes required to form leaves.
B are located at an elevated position on the plant.
C have only the genes for leaf formation switched on.
D suffer loss of root and stem genes during cell division.

20 The following list gives the steps that may be used in the future to culture and make use of human stem cells.

1 Stem cells cloned into colonies in the laboratory.
2 Differentiated cells used to repair damaged organs.
3 Stem cells induced by chemical means to differentiate.
4 Undifferentiated cells extracted from embryo.

Which of the following is the correct sequence of the steps?

A 1, 4, 2, 3
B 4, 1, 3, 2
C 1, 4, 3, 2
D 4, 3, 1, 2

30 Hormonal influences on growth

Match the terms in list X with their descriptions in list Y.

list X		list Y	
1	α-amylase	**a**	hormone that stimulates overall growth of the human body
2	abscission	**b**	hormone that reverses genetic dwarfism and breaks seed dormancy in plants
3	aleurone layer	**c**	directional growth by a plant organ in response to light from one direction
4	apical dominance	**d**	starch-digesting enzyme produced in cereal grains
5	dwarf variety	**e**	chemical messenger from the pituitary gland which controls the activity of the thyroid gland
6	gibberellic acid	**f**	hormone that regulates the human body's metabolic processes
7	herbicide	**g**	separation of a leaf or fruit from its plant of origin
8	indole acetic acid	**h**	strain of plant unable to manufacture sufficient gibberellic acid to attain normal height
9	phototropism	**i**	selective weedkiller often containing synthetic auxin
10	somatotrophin	**j**	inhibition of lateral buds by a high concentration of auxin from the shoot tip
11	thyroid-stimulating hormone	**k**	site of induction of α-amylase production by gibberellic acid in a cereal grain
12	thyroxine	**l**	type of auxin which promotes cell division and elongation in plants

Choose the ONE correct answer to each of the following multiple choice questions.

1 The numbered structures in the diagram represent two human endocrine glands.

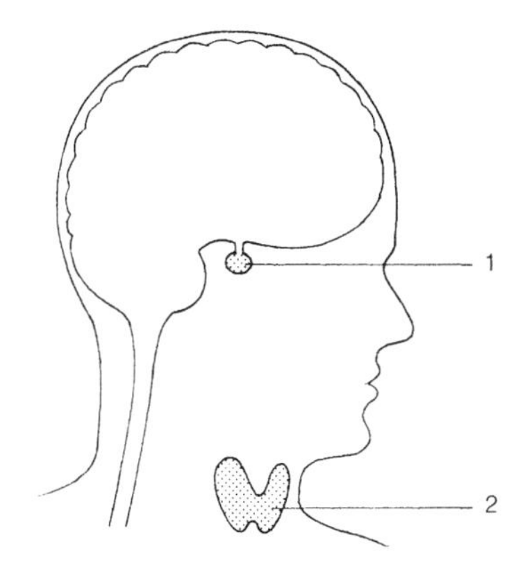

Which line in the table gives the correct sites of production of the three hormones?

	thyroxine	*somatotrophin*	*thyroid-stimulating hormone*
A	2	1	2
B	2	1	1
C	1	2	1
D	1	2	2

Items 2 and 3 refer to the following bar graph.

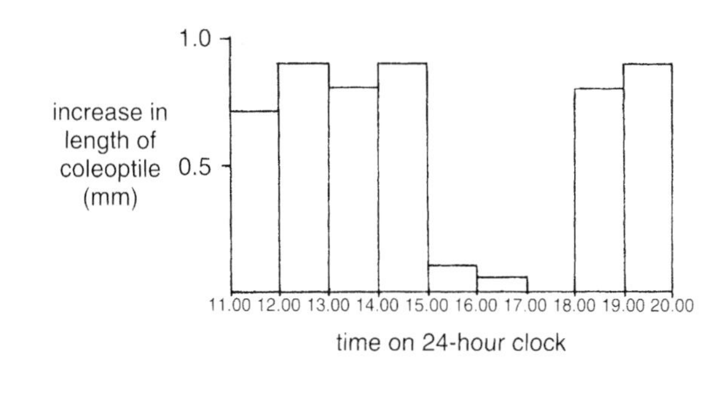

2 The oat coleoptile was decapitated at

A 13.05 hours **B** 14.05 hours
C 15.05 hours **D** 16.05 hours

3 Which of the following shows what was done to the coleoptile stump at 18.05 hours?

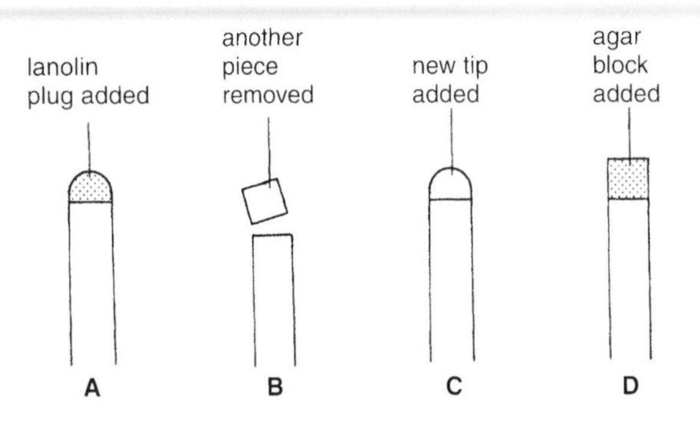

Items 4, 5 and 6 refer to the following graph.

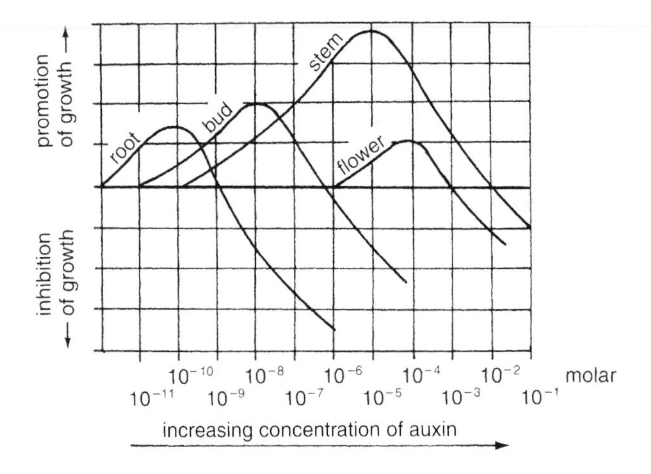

4 10^{-5} molar concentration of auxin causes maximum growth of the

A bud. **B** root. **C** flower. **D** stem.

5 Which molar concentration of auxin inhibits growth of roots yet promotes growth of both buds and stems?

A 10^{-10} **B** 10^{-9} **C** 10^{-8} **D** 10^{-6}

6 Both promotion of stem growth and inhibition of flower growth occur between molar concentrations

A 10^{-5} and 10^{-4}. **B** 10^{-4} and 10^{-3}.
C 10^{-3} and 10^{-2}. **D** 10^{-2} and 10^{-1}.

7 Curvature of a shoot towards a light source is brought about by

A increased plasticity of cell walls on the light side.
B reduced concentration of growth substance on the dark side.
C complete inhibition of cell division on the light side.
D differential elongation of cells behind the shoot tip.

8 The following diagram shows four coleoptiles set up at the start of an experiment.

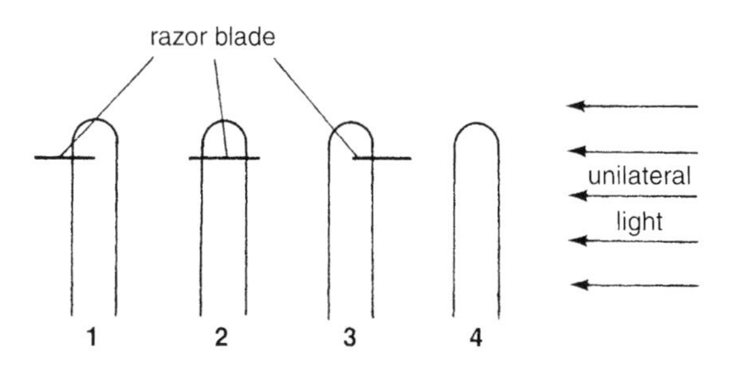

Which two coleoptiles will BOTH bend towards the light source?

A 1 and 2 **B** 1 and 4

C 2 and 3 **D** 3 and 4

9 Agar blocks 1 and 2 were kept in the positions shown in the diagram below for several hours and then transferred onto two freshly cut coleoptiles.

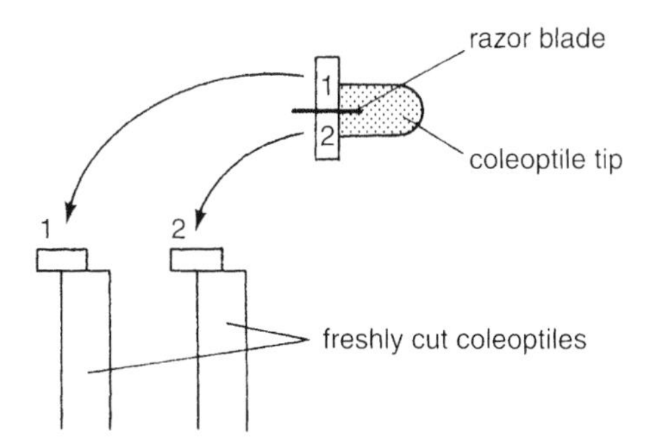

Which of the following would result after two days of growth?

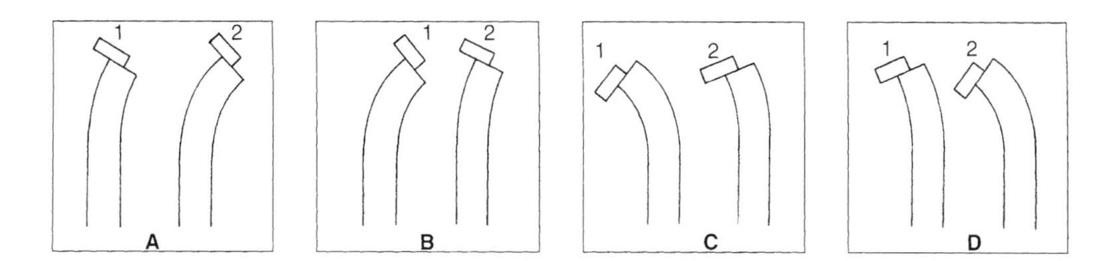

Items **10** and **11** refer to the experiment shown in the following diagram.

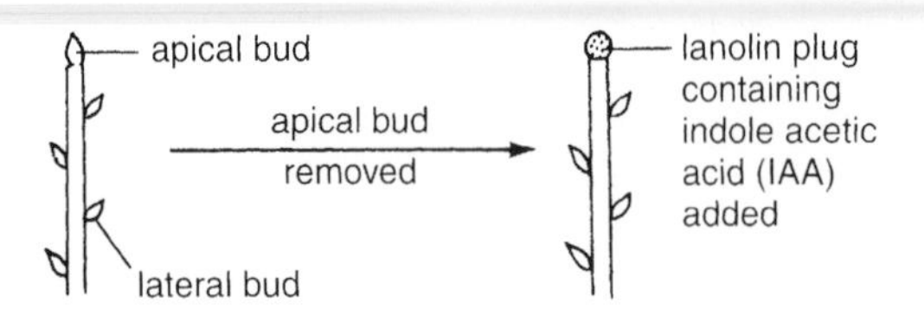

10 After two weeks the appearance of the shoot would be

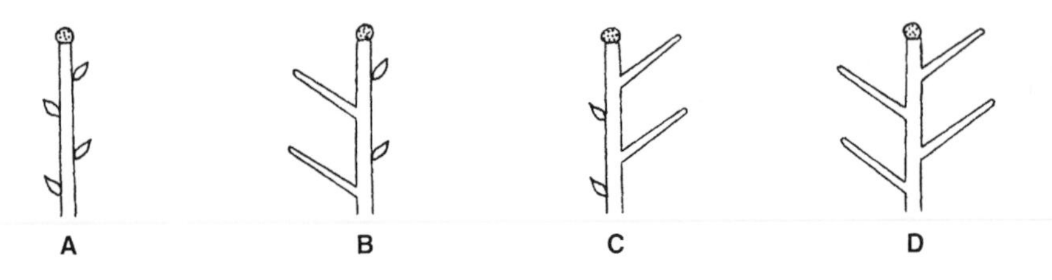

11 The most suitable control for this experiment would be

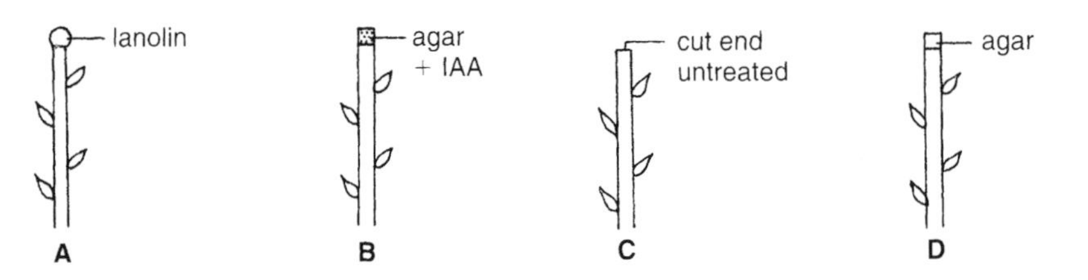

12 Synthetic auxins are NOT used to
A selectively kill broad-leaved garden weeds.
B promote abscission of fruit from fruit trees.
C stimulate formation of roots at the cut ends of stems.
D induce fruit development in unpollinated flowers.

13 The diagram below shows a normal untreated dwarf pea plant.

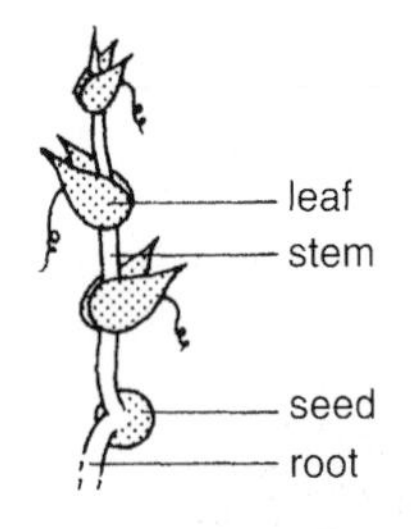

If its shoot had been treated with gibberellic acid during the early stages of germination, it would instead have the appearance

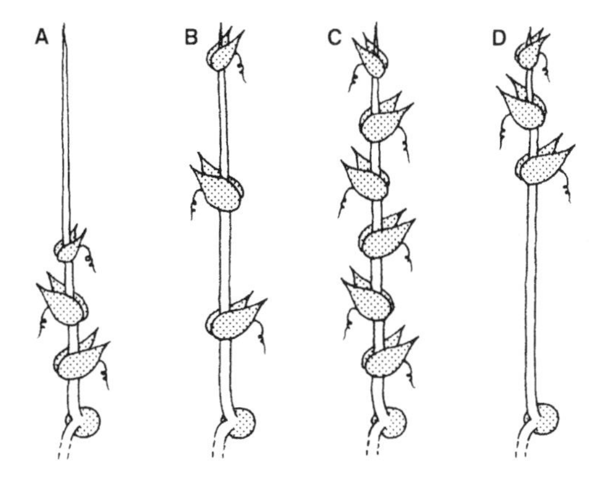

Items **14**, **15** and **16** refer to the following experiment where soaked barley grains are cut and placed, cut surface down, on a plate of starch agar containing gibberellic acid.

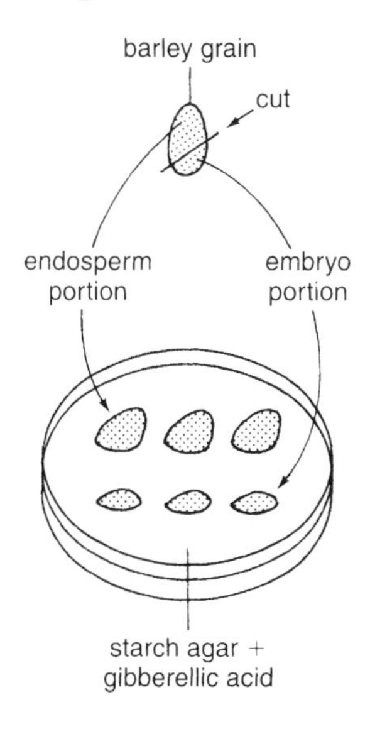

14 Which of the following best represents the appearance of the plate 24 hours later when the grains are removed and the plate is flooded with iodine solution?

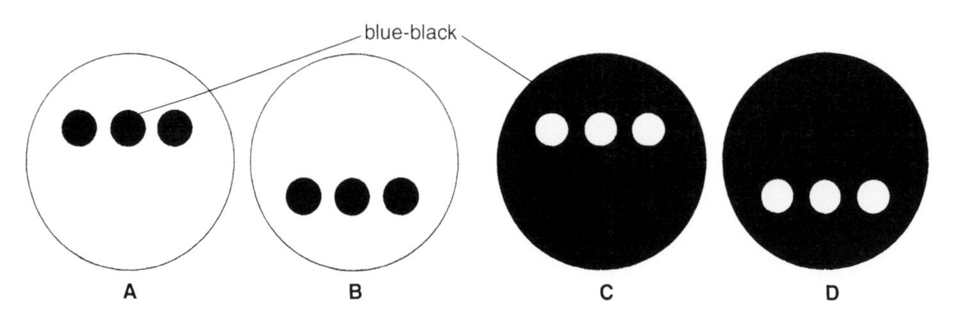

Items **15** and **16** refer, in addition, to the following diagram of the internal structure of a barley grain.

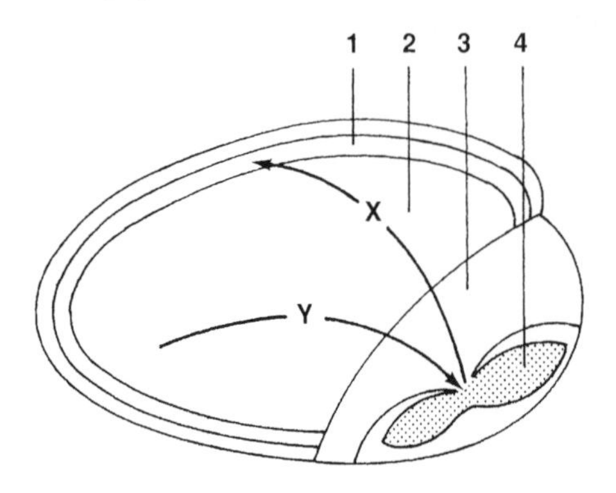

15 Which number indicates the aleurone layer?

A 1 **B** 2 **C** 3 **D** 4

16 It is thought that in a germinating barley grain, gibberellic acid passes along route

A X and digests starchy endosperm to sugar in region 1.

B X and induces α-amylase production in region 1.

C Y and supplies sugar to the cells in region 4.

D Y and induces α-amylase production in region 4.

17 Which of the following effects is brought about by gibberellins but NOT by auxins?

A maintenance of dormancy in lateral buds

B inhibition of leaf abscission

C promotion of phototropic responses

D breaking of dormancy in leaf buds

18 Which of the following effects is brought about by BOTH gibberellic acid and indole acetic acid?

A reversal of genetic dwarfism

B promotion of cell elongation

C breaking of dormancy in seeds

D induction of α-amylase in seed grains.

Items **19** and **20** refer to the graph below which shows the effect of the recommended concentration of a selective herbicide on a cereal crop and three different species of broad-leaved weed at different stages of their development.

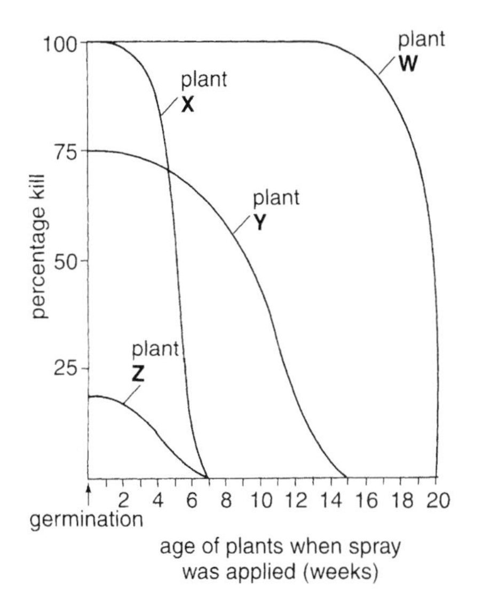

19 Which curve represents the cereal crop?

A W **B** X

C Y **D** Z

20 In which of the following weeks would application of this concentration of the herbicide spray be MOST effective?

A 4 **B** 7

C 15 **D** 20

31 Effects of chemicals on growth

Match the terms in list X with their descriptions in list Y.

list X

1 alcohol

2 calcium

3 iron

4 lead

5 magnesium

6 nicotine

7 nitrogen

8 phosphorus

9 potassium

10 rickets

11 thalidomide

12 vitamin D

list Y

a element needed for the production of ATP

b element which inhibits the activity of certain enzymes

c chemical present in tobacco which retards the growth and mental development of a human foetus

d element needed by a plant for an essential role in membrane transport

e element needed by the human body for the production of haemoglobin

f condition resulting in soft bones caused by insufficient vitamin D in the diet

g chemical which can cause limb deformation of a human fetus during development

h element needed by the human body for the normal growth of bones and teeth

i chemical needed by humans to promote the uptake of calcium from the small intestine

j chemical present in beer, wine and spirits which can retard physical and mental growth of a developing human fetus

k element needed for the synthesis of amino acids

l element needed by a plant for the synthesis of chlorophyll

Choose the ONE correct answer to each of the following multiple choice questions.

1 Which of the following is NOT required when setting up a water culture experiment (to determine the importance of individual chemical elements to a plant)?
A distilled water containing one essential element
B a glass tube to aerate the plant's roots
C nitric acid to remove mineral traces from glassware
D an opaque cover around the culture solution

Items 2, 3 and 4 refer to the chemical elements given in the following list.
1 potassium **2** nitrogen **3** magnesium
4 phosphorus **5** iron

2 Which element is an essential constituent of all proteins?
A 1 **B** 2 **C** 3 **D** 4

3 The formation of chlorotic leaves is the result of a plant being deficient in EITHER
A 1 or 2. **B** 2 or 3. **C** 3 or 4. **D** 1 or 3.

4 Which element is required by a plant for the formation of ATP and nucleic acids?
A 1 **B** 3 **C** 4 **D** 5

5 The following table refers to four test tubes set up to investigate the effect of catechol oxidase (an enzyme present in apple cells) on catechol, its substrate, at room temperature.

tube	*catechol*	*state of apple tissue*	*lead nitrate*
1	✓	fresh	✓
2	✓	boiled	✓
3	✓	boiled	✗
4	✓	fresh	✗

(✓ = present, ✗ = absent)

Within a few minutes, the apple tissue would become dark brown in
A tube 1 only. **B** tubes 1 and 2.
C tubes 3 and 4. **D** tube 4 only.

Items **6** and **7** refer to the table below which gives the results of a survey on lead in drinking water.

	number of persons				
lead concentration in drinking water (mg/l) / *lead concentration in blood (mg/l)*	*0.000 to 0.010*	*0.011 to 0.050*	*0.051 to 0.100*	*0.101 to 0.300*	*0.301 to above*
0.000 to 0.100	15	2	0	1	0
0.101 to 0.200	27	10	9	11	3
0.201 to 0.300	8	3	2	10	7
0.301 to 0.400	0	0	1	3	4
0.401 to 0.500	0	0	0	1	2
0.501 and above	0	0	0	0	1

6 The most common concentration of lead present in drinking water and the number of people found consuming it are respectively

	lead concentration (mg/l)	*number of persons*
A	0.000 to 0.010	50
B	0.000 to 0.010	60
C	0.101 to 0.200	50
D	0.101 to 0.200	60

7 The highest concentration of lead found in blood and the number of people possessing it are respectively

	lead concentration (mg/l)	*number of persons*
A	0.301 and above	1
B	0.301 and above	17
C	0.501 and above	1
D	0.501 and above	17

8 Iron is NOT lost from the human body in

A bile. **B** urine.
C saliva. **D** dead skin cells.

9 Iron is an essential ingredient of BOTH
A haemoglobin and cytochrome.
B cytochrome and lymph.
C lymph and plasma.
D plasma and haemoglobin.

10 Calcium is NOT required by the human body for
A clotting of blood.
B formation of teeth.
C contraction of muscle.
D manufacture of bile.

11 The table shows the calcium content (per 100 g) of several foodstuffs.

foodstuff	*calcium content (mg/100 g)*
almonds	250
blackcurrants	60
chocolate biscuits	130
cottage cheese	80
herring	100
ice cream	140
milk	120
milk chocolate	250
oranges	40
sardines	400
white bread	100
yoghurt	140

The minimum recommended daily intake of calcium for a 16 year old is 600 mg. This could be achieved by eating
A 100 g almonds, 200 g oranges and 200 g chocolate biscuits.
B 100 g cottage cheese, 200 g ice cream and 400 g blackcurrants.
C 100 g milk chocolate, 200 g herring and 100 g yoghurt.
D 50 g sardines, 250 g white bread and 100 g milk.

12 In growing children, vitamin D is essential to promote
A reabsorption of minerals from the bones.
B storage of excess iron in the liver.
C uptake of iodine by the thyroid gland.
D absorption of calcium from the intestine.

13 Which of the following foods are BOTH rich in vitamin D?

A cod-liver oil and blackcurrants
B blackcurrants and raw carrots
C egg yolk and cod-liver oil
D raw carrots and egg yolk

14 If a pregnant woman drinks alcohol in excess, the developing embryo is adversely affected because

A it fails to receive an adequate oxygen supply.
B its nervous system is artificially stimulated.
C it receives blood containing carbon monoxide.
D it is deprived of soluble carbohydrate.

15 Which column in the following table refers to the possible effects of thalidomide on a developing human fetus?

possible effect	A	B	C	D
retardation of general physical growth	✓	✓		
malformation of eyes and ears			✓	✓
heart defects	✓		✓	✓
failure of limbs to develop				✓
painful withdrawal symptoms suffered after birth		✓		

32

Effect of light on growth

Match the terms in list X with their descriptions in list Y.

list X	list Y
1 day neutral plant	**a** growth movement by a plant in response to a unidirectional light source
2 etiolation	**b** plant which only flowers when the daylength is below a certain critical level
3 long day breeder	**c** plant in which flowering is not dependent upon photoperiod
4 long day plant	**d** animal whose reproductive activity is stimulated by increasing daylengths in spring
5 photoperiodism	**e** development of long internodes and small yellow leaves in plants deprived of light
6 phototropism	**f** animal whose reproductive activity is stimulated by decreasing daylengths in autumn
7 short day breeder	**g** plant which only flowers when the daylength is above a certain minimum length
8 short day plant	**h** response by a living organism to a photoperiod of a certain length

Choose the ONE correct answer to each of the following multiple choice questions.

1 The table below shows a comparison between two genetically identical plants, one grown in dark, the other in light.

Which paired statement is INCORRECT?

	in dark	*in light*
A	leaves remain curled	leaves become expanded
B	weak internodes develop	strong internodes develop
C	leaves fail to make chlorophyll	leaves make chlorophyll
D	internodes remain short	elongated internodes develop

2 A short day plant will flower only when the continuous period of

A light is above a critical level.
B light is below a critical level.
C darkness is below a critical level.
D darkness is interrupted at a critical level.

3 Sedum is a long day plant. Its critical duration of light is 13 hours. Under which of the following conditions would it flower?

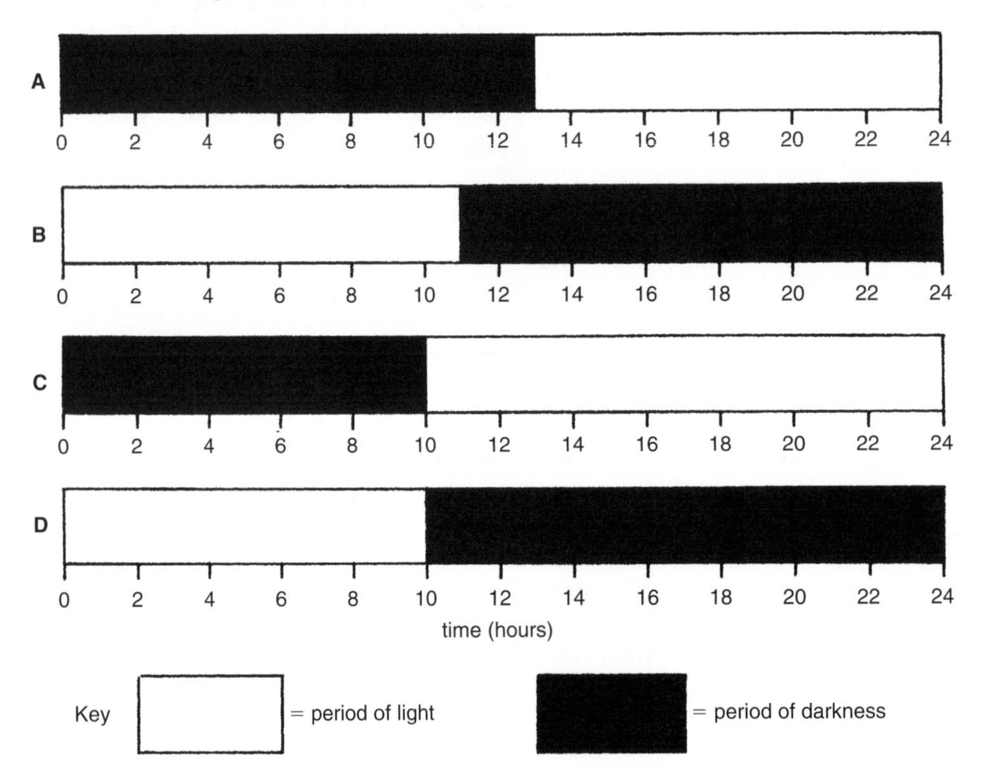

4 Maryland Mammoth Tobacco is a short day plant. Its critical duration of darkness is 10 hours. Under which of the following conditions will it NOT flower?

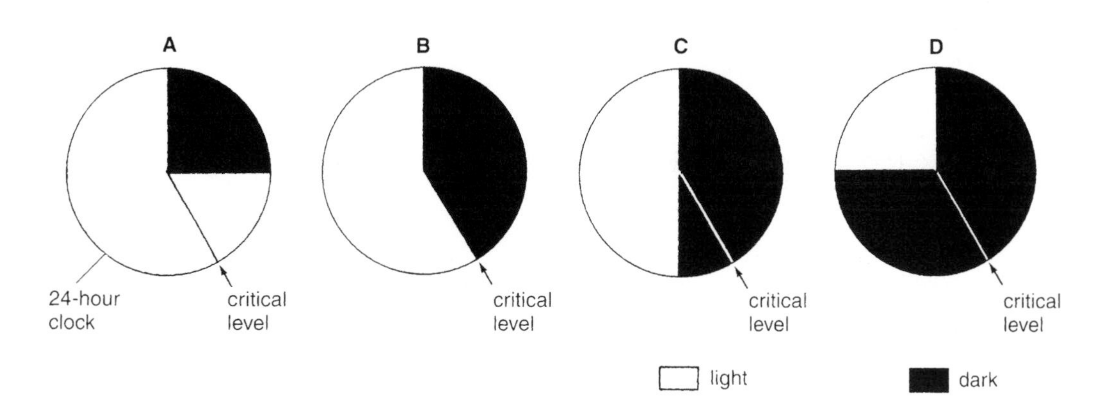

5 When a certain species of flowering plant is exposed to artificially controlled periods of light and dark, it responds as shown in the following table.

dark period (hours)	*light period (hours)*	*flowering (+) or no flowering (–)*
10	12	+
11	11	–
11	12	+
12	11	–
12	12	+
13	11	–

This species is a

A long day plant and the critical duration of light is 11 hours.
B long day plant and the critical duration of light is 12 hours.
C short day plant and the critical duration of darkness is 11 hours.
D short day plant and the critical duration of darkness is 12 hours.

6 Cocklebur is a short day plant. After exposure to the treatments described in the following diagram for several days, plants X and Y were grafted together.

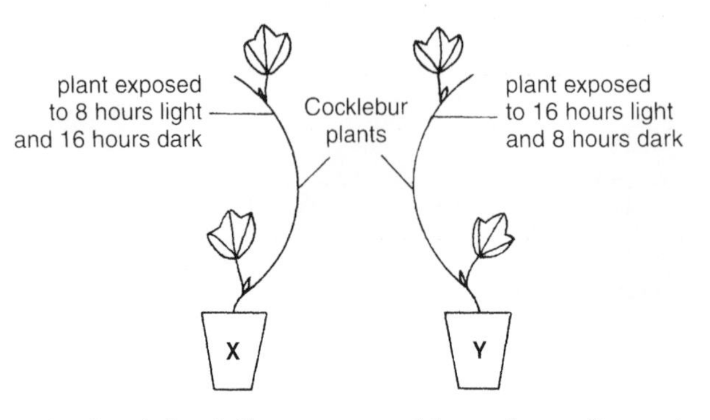

Which of the following would result on flowering?

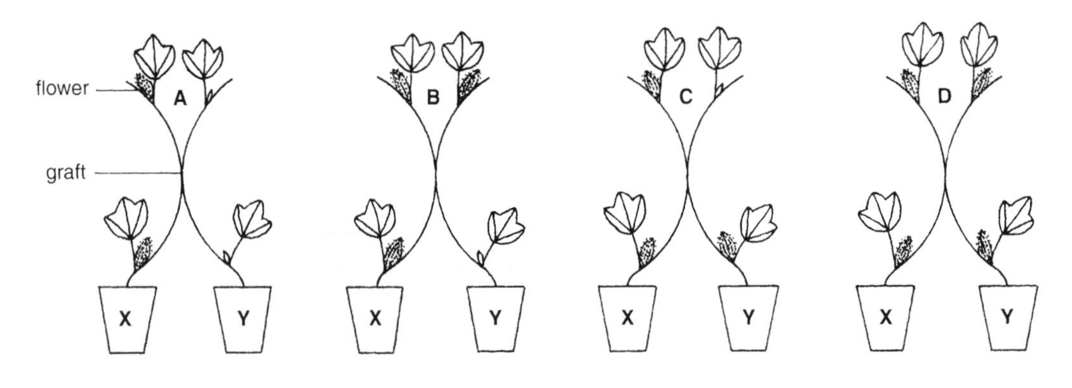

7 Some of the events which precede mating behaviour in sheep are

1 stimulation of the hypothalamus
2 secretion of hormones which induce mating behaviour
3 arrival of shorter photoperiods
4 stimulation of the pituitary gland

The order in which these events occur is

A 3, 1, 4, 2. **B** 3, 4, 1, 2. **C** 1, 3, 4, 2. **D** 1, 3, 2, 4.

Items 8 and 9 refer to the following graph and table which refer to testis volume of the wood pigeon and changing photoperiod, respectively.

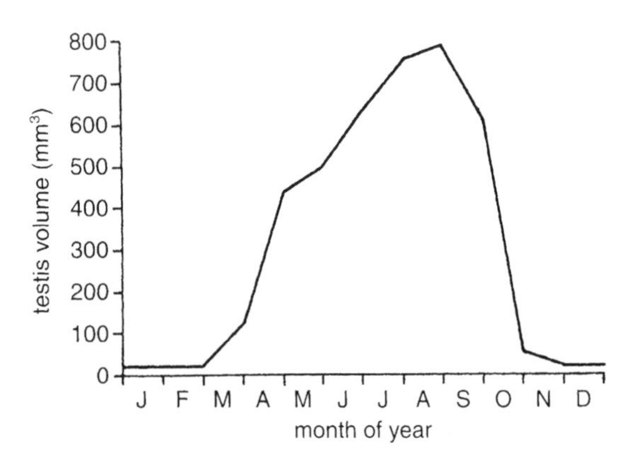

month of year	*average photoperiod (hours)*
J	8
F	10
M	12
A	14
M	16
J	18
J	17
A	15
S	13
O	11
N	9
D	7

8 If breeding behaviour and release of sperm do not occur until the testis has reached a volume of 300 mm^3, then the critical photoperiod (in hours) is
A 12. **B** 14. **C** 16. **D** 18.

9 The breeding season for this bird is from
A March to September. **B** April to September.
C March to October. **D** April to October.

10 The table below contains data obtained from wild populations of three species of bird living in the Northern Hemisphere.

species	*latitude (°)*	*maximum gonad size (mm^3)*
X	34	400
X	45	490
Y	47	700
Y	52	1500
Z	52	450
Z	57	955

Which of the following statements is CORRECT?

A Compared with X and Y, gonad size is greatest in species Z.

B For each species, increase in latitude is accompanied by an increase in gonad size of at least 50%.

C The higher the latitude, the greater the gonad size of all three species.

D An increase in latitude from 34° to 57° is accompanied by an increase in gonad size of 555 mm^3 in species X.

33

Physiological homeostasis

Match the terms in list X with their descriptions in list Y.

list X		list Y	
1	adrenaline	**a**	region of the brain containing a centre which regulates body temperature
2	anti-diuretic hormone	**b**	process by which the bore of skin arterioles becomes narrower
3	diabetes mellitus	**c**	hormone that promotes the conversion of glycogen to glucose but does not affect heart rate
4	ectotherm	**d**	structure that detects changes in the blood's water concentration
5	endotherm	**e**	maintenance of the body's water balance at a normal level
6	glucagon	**f**	process by which the bore of skin arterioles becomes wider
7	hypothalamus	**g**	organ that produces insulin and glucagon
8	insulin	**h**	structure which releases ADH into the bloodstream
9	internal environment	**i**	chemical messenger which brings about osmoregulation by negative feedback control
10	liver	**j**	structure which detects changes in body temperature
11	osmoreceptors	**k**	condition caused by lack of insulin production
12	osmoregulation	**l**	hormone that promotes the conversion of glucose to glycogen
13	pancreas	**m**	hormone that speeds up heart rate and promotes the breakdown of glycogen to glucose
14	physiological homeostasis	**n**	animal which is unable to regulate its body temperature by physiological means
15	pituitary gland	**o**	human body's community of cells and the tissue fluid that bathes them
16	thermoreceptor	**p**	site of glycogen storage in the human body
17	vasoconstriction	**q**	animal which is able to regulate its body temperature by physiological means
18	vasodilation	**r**	maintenance of the body's internal environment within tolerable limits by negative feedback control

Choose the ONE correct answer to each of the following multiple choice questions.

1 Negative feedback control involves the following four stages
1 effectors bring about corrective responses
2 a receptor detects a change in the internal environment
3 deviation from the norm is counteracted
4 nerve or hormonal messages are sent to effectors

The order in which these occur is
A 2, 1, 4, 3. **B** 2, 4, 1, 3. **C** 2, 4, 3, 1. **D** 4, 2, 1, 3.

2 In humans, an increase in urine production occurs as a result of
A a decrease in environmental temperature.
B a decrease in water uptake.
C an increase in salt uptake.
D an increase in rate of sweating.

3 When a decrease in water concentration of the blood occurs, which of the following series of events brings about homeostatic control?

	ADH production	*permeability of kidney tubules*	*volume of urine produced*
A	increased	increased	decreased
B	increased	decreased	decreased
C	decreased	increased	decreased
D	increased	increased	increased

4 Set point is defined as the
A point at which the body corrects any change in internal environment.
B range within which the level of all factors must remain.
C optimum position about which any one factor varies continuously.
D point at which the body always maintains the internal environment.

5 The accompanying diagram shows a simplified version of the homeostatic control of blood sugar level in the human body.

Which of the following responses would occur in a human body as a direct result of eating a school lunch?
A R, P and T **B** S, P and T
C R, Q and T **D** R, P and U

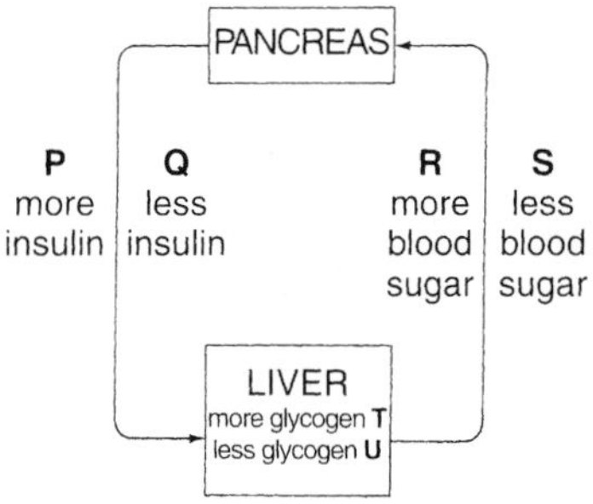

Items **6**, **7** and **8** refer to the diagram where the four lettered structures represent endocrine glands.

6 Which is the site of glucagon production?

7 Which is the site of anti-diuretic hormone production?

8 Which of these glands makes adrenaline?

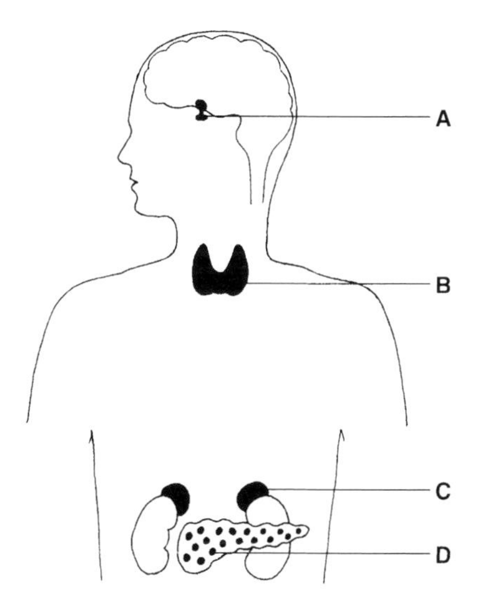

Items **9** and **10** refer to the following graph which shows the blood sugar level of a person who has consumed 50 g of glucose at the time indicated.

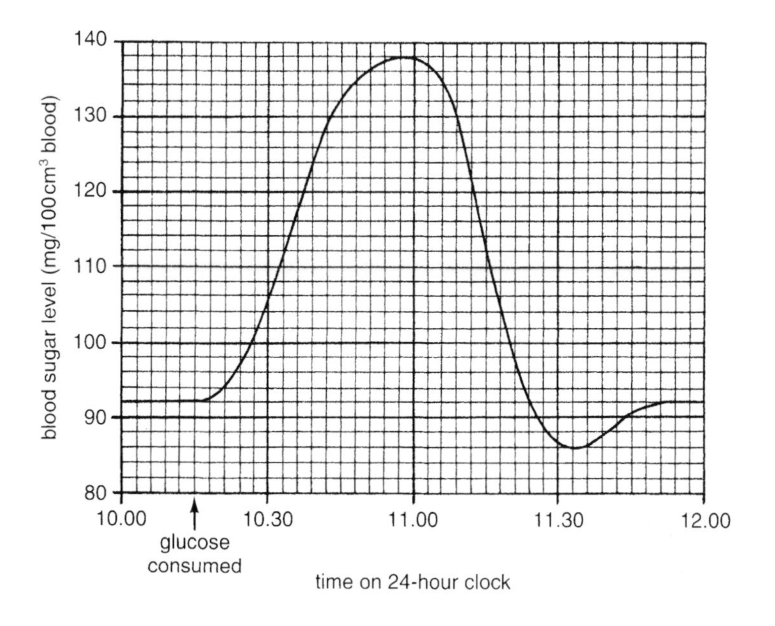

9 The maximum level of blood sugar expressed as a multiple of the normal (resting) value is

A 0.5. **B** 0.67. **C** 1.5. **D** 12 696.

10 Compared with normal levels, extra insulin and extra glucagon would be present at certain times during this two-hour period. Two of these times are:

	extra insulin	*extra glucagon*
A	10.30	11.00
B	10.30	11.30
C	11.00	10.30
D	11.30	10.30

11 The temperature-monitoring centre in the human brain is situated in the

A cerebellum. **B** pituitary.
C medulla. **D** hypothalamus.

12 Following overheating of a mammal's internal environment, the skin acts as an effector and the following changes occur.

	altered state of arterioles leading to skin	*altered state of erector muscles in skin*
A	constricted	relaxed
B	dilated	relaxed
C	dilated	contracted
D	constricted	contracted

Items **13**, **14** and **15** refer to the following diagram.

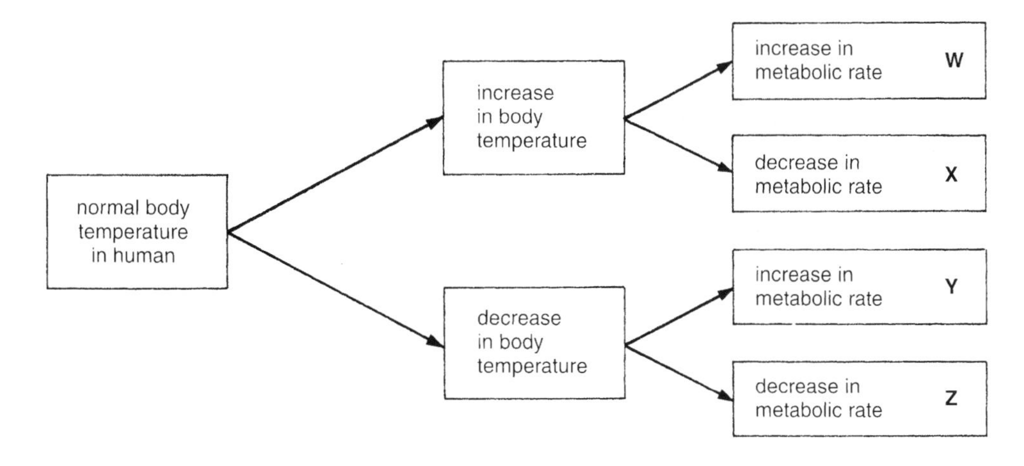

13 Which letters indicate the normal negative feedback control of body temperature?

A W and Y **B** W and Z
C X and Y **D** X and Z

14 Which situation would be the immediate result of exposure to intense cold?

A W **B** X **C** Y **D** Z

15 Which situation would be the result of prolonged exposure to intense cold leading to hypothermia?

A W **B** X **C** Y **D** Z

Items **16** and **17** refer to the following possible answers.

A whale and herring **B** herring and shark
C shark and dolphin **D** dolphin and whale

16 Which of these animals are BOTH endotherms?

17 Which of these animals are BOTH ectotherms?

18 Bacteria taken from the intestine of a certain animal were found to grow on a culture plate at 7°C but not at 37°C. From which animal had they come?

A cod **B** seal **C** penguin **D** polar bear

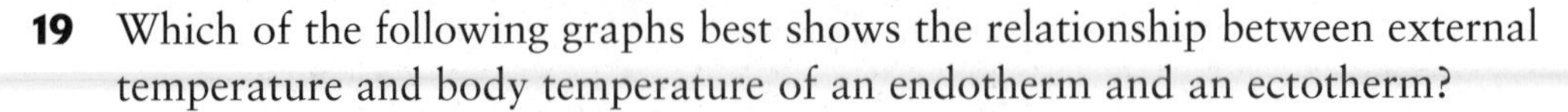

19 Which of the following graphs best shows the relationship between external temperature and body temperature of an endotherm and an ectotherm?

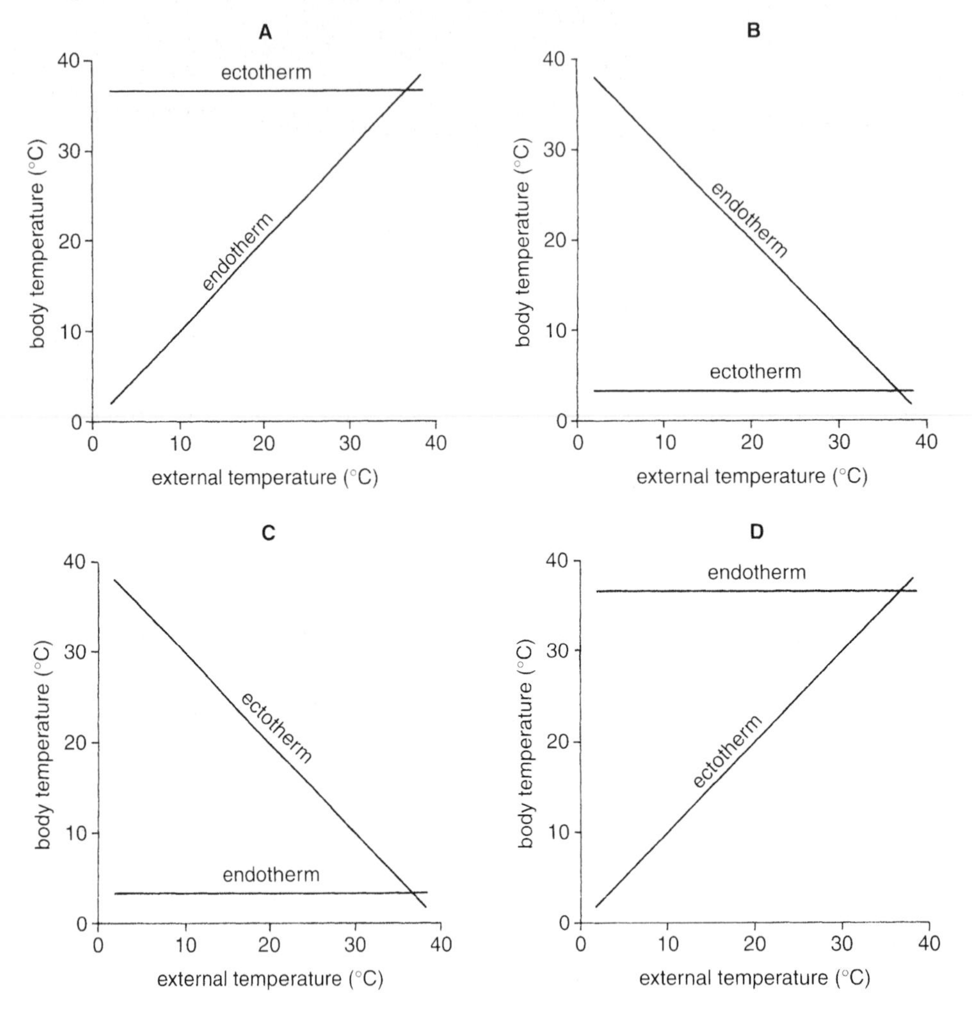

20 The responses to external environment shown in the following table are typical of one of the animals listed below the table. Which one?

temperature of of habitat (°C)	*activity of animal*	*metabolic rate*
0	inactive	low
10	inactive	low
20	active	normal
30	very active	high

A gerbil **B** lizard **C** vulture **D** camel

34 Regulation of populations

Match the terms in list X with their descriptions in list Y.

list X

1 birth rate

2 carrying capacity

3 death rate

4 density-dependent

5 density-independent

6 dynamic equilibrium

7 environmental resistance

8 homeostasis

9 population

10 population density

11 population dynamics

list Y

a set of factors that affect the size of a population and prevent it from increasing in size indefinitely

b study of population changes and the factors which bring them about

c steady state shown by a population that has reached the carrying capacity of its environment

d group of individuals of the same species

e negative feedback control that regulates population size

f the number of individuals present per unit area or volume of a habitat

g measure of the number of individuals within a population that died during a certain interval of time

h measure of the number of new individuals produced by a population during a certain interval of time

i term referring to a factor that affects growth of a population regardless of the population's density

j term referring to a factor that only affects a population once it has reached a certain size

k maximum size of a population that can be maintained by the resources available in an environment

Choose the ONE correct answer to each of the following multiple choice questions.

Items **1** and **2** refer to the following graph which plots the birth and death rates for a certain country over a period of 80 years.

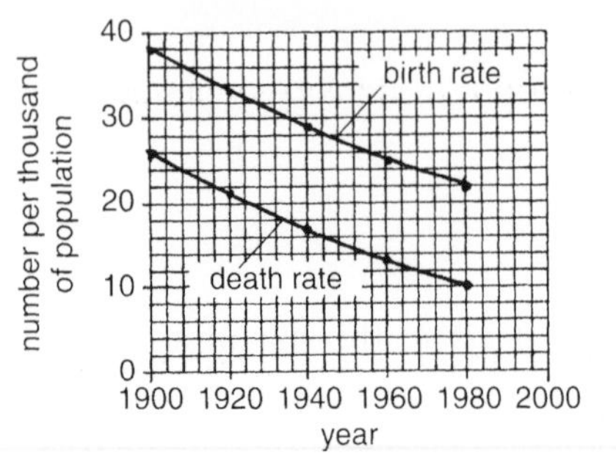

1 What was the PERCENTAGE annual increase in population during this period of time?

A 0.6 **B** 1.2 **C** 6 **D** 12

2 After 1980 the population continued to increase in size but showed a decrease in RATE of growth. Which of the following graphs best represents this trend?

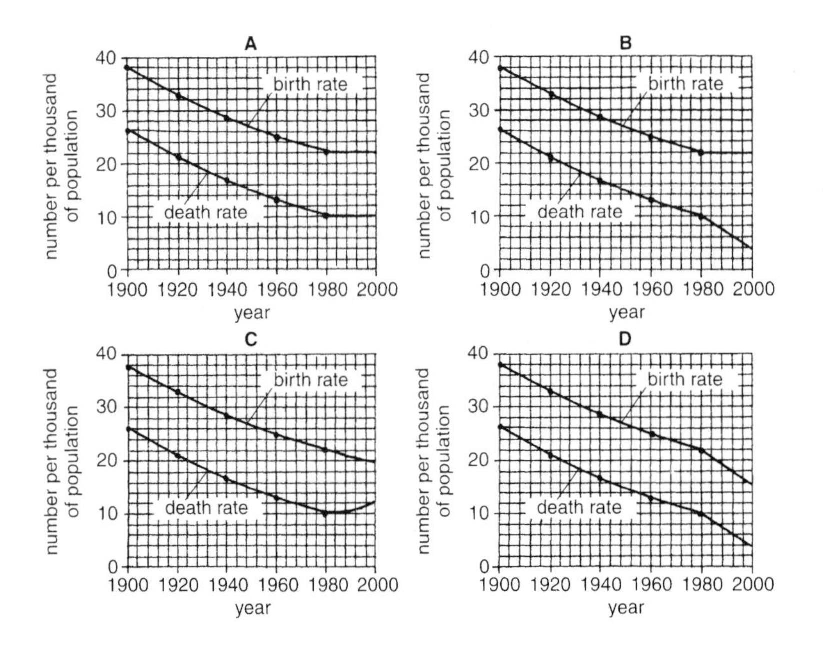

Items **3** and **4** refer to the following possible answers.

A dynamic equilibrium **B** carrying capacity

C population dynamics **D** environmental resistance

3 Which term means the upper limit in size of a population that can be maintained by the resources available in its ecosystem?

4 Which expression refers collectively to those factors that prevent a population from increasing in size indefinitely?

5 Which of the following factors acts on a population in a density-dependent manner?

A drought | **B** pollution
C flooding | **D** predation

6 Which of the following acts as a density-independent factor on a population?

A disease | **B** shortage of food
C forest fire | **D** excretory wastes

7 Which of the following graphs shows the effect of a density-dependent factor on a population?

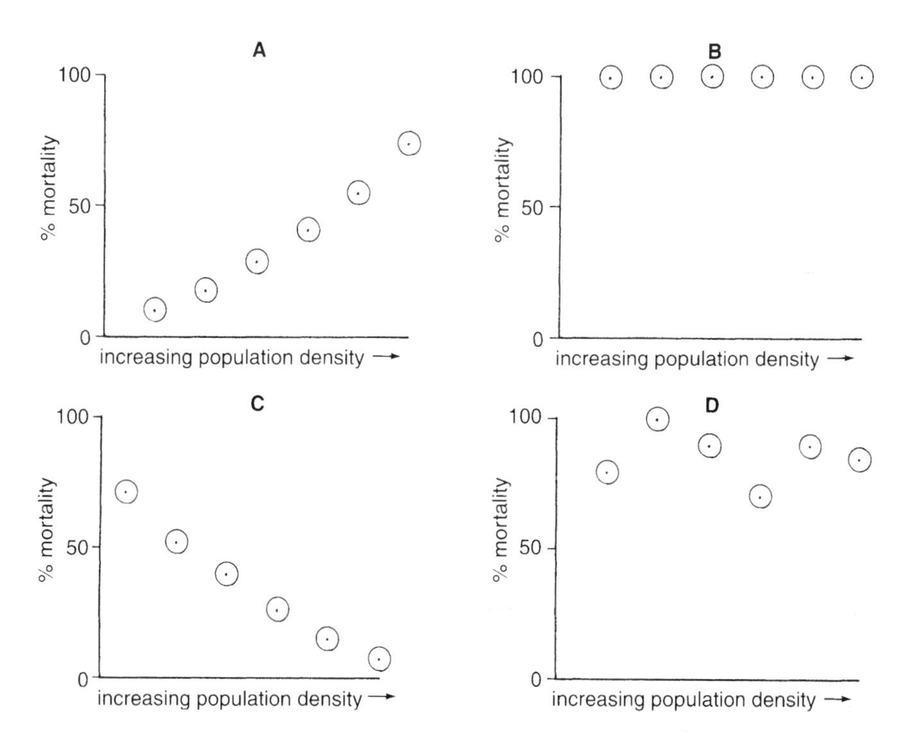

8 The data in the following table refer to the results from a series of experiments using water fleas.

population density of adults (number/litre of pond water)	*average number of offspring produced per day by female aged:*		
	10 days	*15 days*	*20 days*
1000	2.5	3.2	3.7
4000	1.6	1.9	2.1
8000	0.7	1.1	1.3
16 000	0.4	0.5	0.7

From these data it can be concluded that

A birth rate falls with increased crowding which acts in a density-dependent manner.

B birth rate rises with increased crowding which acts in a density-dependent manner.

C birth rate falls with increased crowding which acts in a density-independent manner.

D birth rate rises with increased crowding which acts in a density-independent manner.

9 Which of the following graphs BEST represents a typical predator-prey relationship?

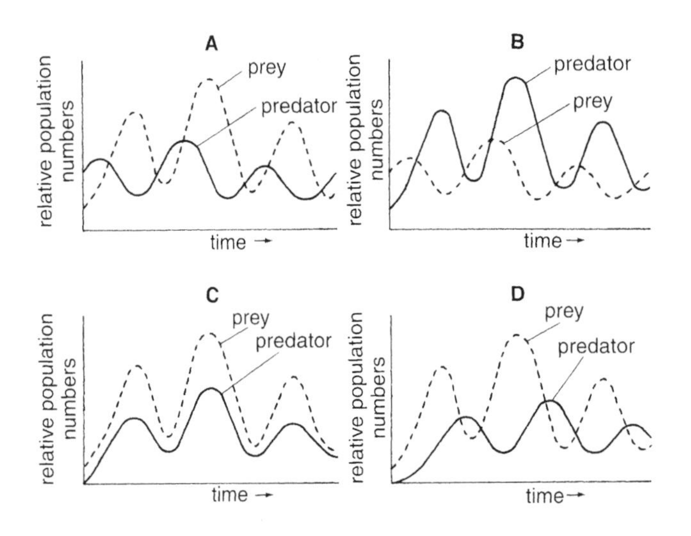

10 Which of the following diagrams correctly represents homeostatic control of population size?

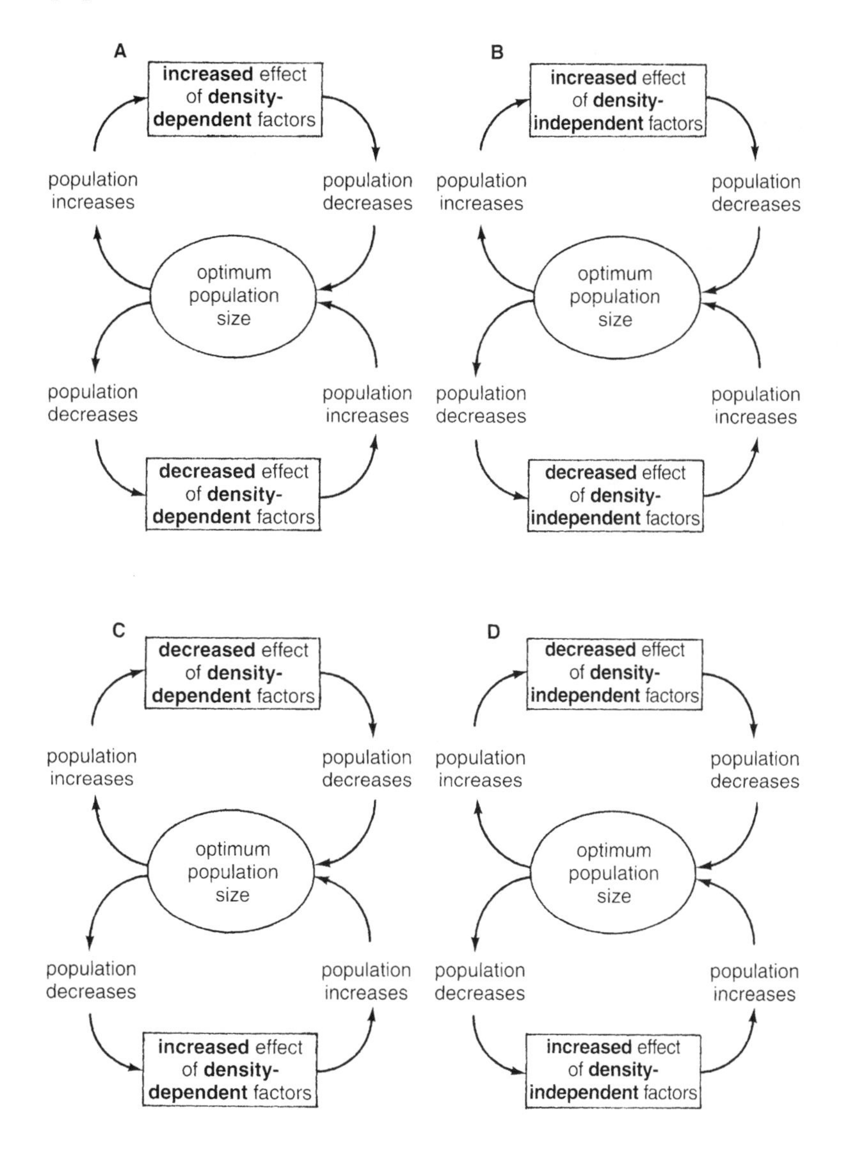

Items **11** to **13** refer to the following diagram which charts the results from an experiment to investigate the effect of soil pH on the distribution of four plant species (W, X, Y and Z).

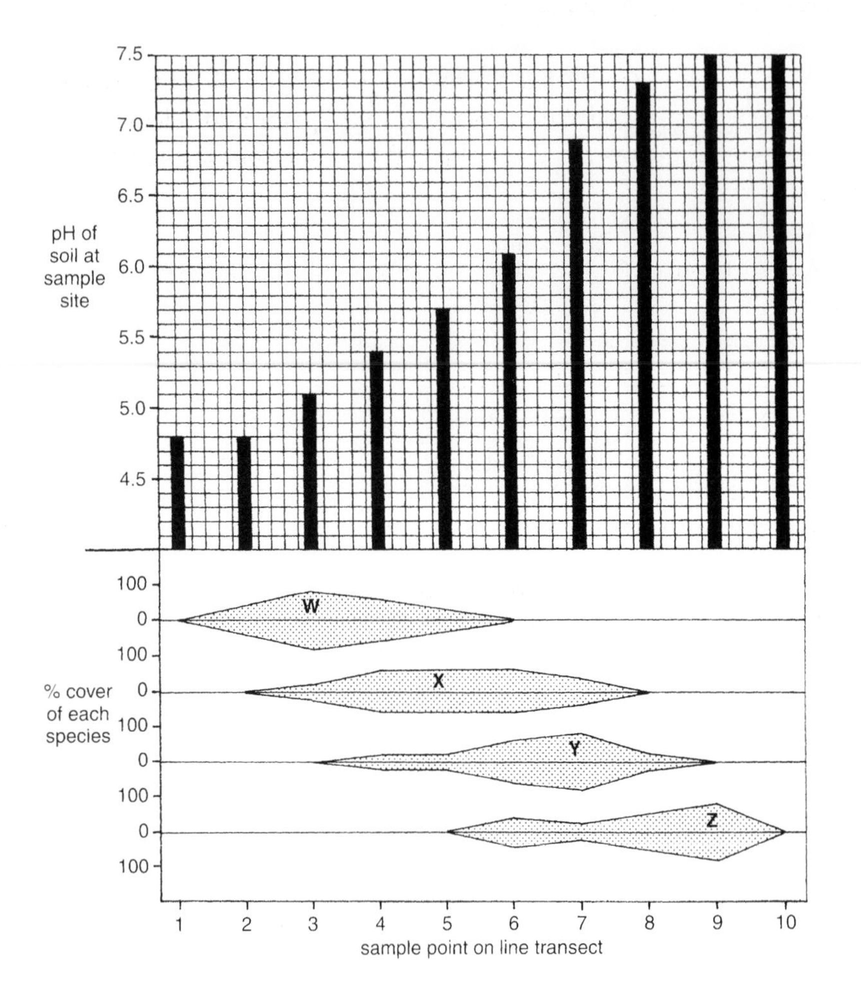

The soil pH and percentage cover of each plant species were recorded at each of 10 sample points at regular intervals along a line transect. The plant cover data are presented as kite diagrams (i.e. symmetrical line graphs on either side of a base line).

11 Only two of the plant species were recorded at sample point
A 3. **B** 4. **C** 5. **D** 6.

12 Species W, X and Y were all found to be growing in soil of pH
A 5.1. **B** 5.7. **C** 6.1. **D** 6.9.

13 Which species is able to tolerate the widest range of soil pH?
A W **B** X **C** Y **D** Z

Items **14** and **15** refer to the following information.

A biologist wished to estimate the total population number of a species of moth living in a wood. She captured 40 moths, marked them with non toxic paint and released them. The next night she set up a light trap and caught 100 specimens of the moth. Of these, eight bore the paint mark.

14 Based on these figures, the total population number of this species of moth is
A 200. **B** 320. **C** 500. **D** 4000.

15 If m^1 = number caught and marked on the first night,
y = total captured on the second night,
m^2 = number captured on the second night bearing the paint mark
and x = total number in the population,
then a simple formula for the estimation of x is

A $\frac{x}{m^2} = \frac{y}{m^1}$ **B** $\frac{m^1}{x} = \frac{y}{m^2}$ **C** $\frac{m^2}{x} = \frac{m^1}{y}$ **D** $\frac{x}{m^1} = \frac{y}{m^2}$

35 Monitoring populations

Match the terms in list X with their descriptions in list Y.

list X		list Y	
1	algal bloom	**a**	agreed suspension of an activity such as whaling
2	culling	**b**	microscopic aquatic algae
3	epidemic	**c**	non-biodegradable chemical used to kill pests
4	indicator species	**d**	planned harvesting of a wildlife species
5	lichen	**e**	type of living organism that poses a threat to mankind's health or economy
6	monoculture	**f**	widespread occurrence of a disease
7	moratorium	**g**	organism whose numbers show the state of an environment's health
8	organochlorine	**h**	contamination of an environment by harmful substances
9	pest species	**i**	result of rapid growth of populations of microscopic water plants
10	phytoplankton	**j**	population of one type of plant
11	pollution	**k**	simple plant which indicates the level of air pollution by sulphur dioxide gas

Choose the ONE correct answer to each of the following multiple choice questions.

1 An indicator organism is one which
A is always restricted to life in badly polluted environments.
B provides information about the fixed quotas of a food species.
C shows that the members of a certain species are threatened with extinction.
D can be used to quantify relative levels of pollution in the biosphere.

Items 2 and 3 refer to the accompanying graph which charts the effect of increased intensity of fishing on four species of edible fish caught in the North Sea. (Sustained yield means the maximum yield of fish that can be maintained from year to year.)

2 Which species is LEAST affected by increased intensity of fishing?

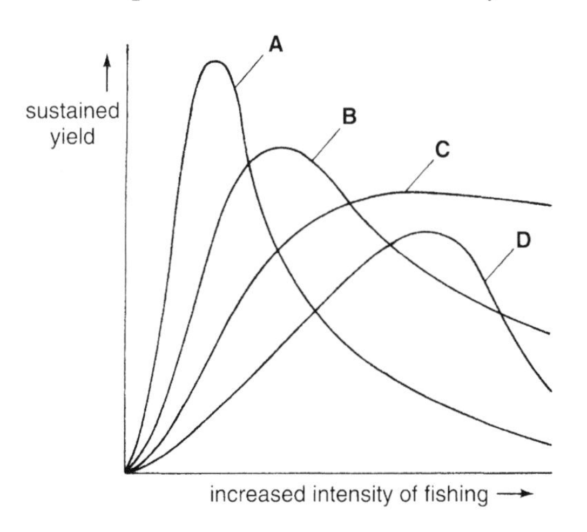

3 Which species of fish is most likely to be the slowest-growing one?

Items 4, 5 and 6 refer to the following table which summarises the results from an investigation into the impact of a new pesticide on four pests.

crop	*pest*	*region of host attacked*	*average loss of crop (acres/year)*	
			without insecticide	*with insecticide*
apple	aphid	leaf and flower	12 000	600
pea	weevil	leaf and pod	4500	450
potato	leather jacket	root and tuber	1800	1700
cabbage	caterpillar	leaf and leaf stalk	3200	400

4 The number of acres of pea crop saved per year by the use of the chemical was
A 450. **B** 4005. **C** 4050. **D** 4500.

5 On which crop did the chemical have the GREATEST effect relative to the others?
A apple **B** pea **C** potato **D** cabbage

6 On which pest was the insecticide LEAST effective?
A aphid **B** weevil **C** leather jacket **D** caterpillar

7 The year in the accompanying graph in which more than 45 regions reported swarms of locusts and an annual rainfall in excess of 75 cm was
A 1975. **B** 1976. **C** 1977. **D** 1978.

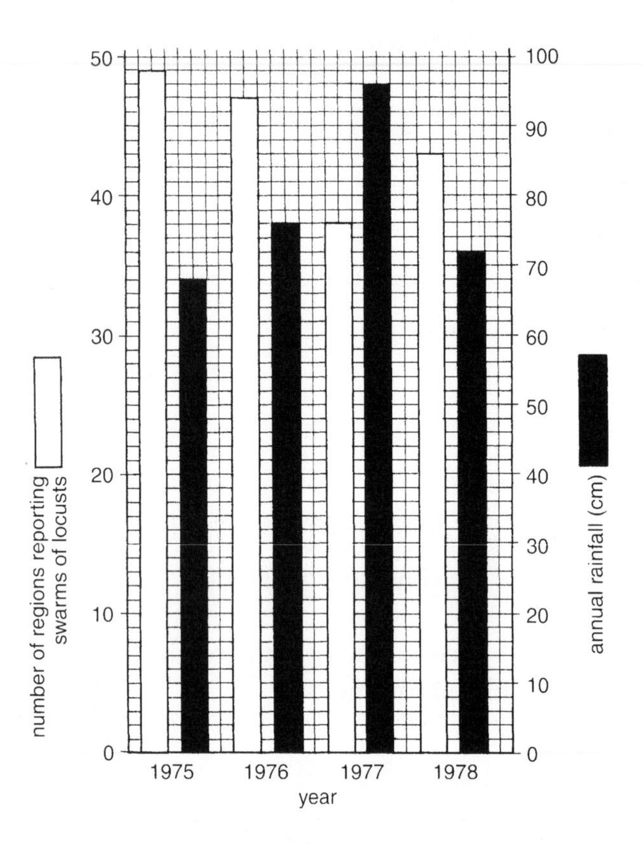

Items **8**, **9** and **10** refer to the following passage.

The spiderwort is a common wild flowering plant found growing in many parts of North America. It has long been known to act as an indicator of environmental degradation caused by excesses of sulphur dioxide,
(5) motor engine exhaust fumes and pesticides. However,

recently it has also revealed itself to be an ultra-
sensitive monitor of ionising radiation.

Humans normally receive about 100 millirems of
background radiation per year. When spiderwort is
(10) exposed to as low a level as 150 millirems, the colour
of the hair cells on its stamens changes from blue to
pink. This occurs as a result of the destruction by
radiation of the genetic material needed to produce
blue pigment.

(15) Whereas most organisms normally take a long time to
show any noticeable effects in response to low (but
unacceptable) levels of radiation, spiderwort is so
sensitive that its colour change occurs within 10–18
days of exposure.

8 Which of the following is the most appropriate title for the passage?
A Useful indicator organisms
B Floral monitor of radiation
C Forms of atmospheric pollution
D Mutation frequency in spiderwort

9 Which of the following is NOT referred to in the passage?
A radioactive fallout
B atmospheric pollution
C excessive use of sprays in agriculture
D effect of organic waste on rivers

10 Which numbered lines in the passage contain information that verifies that radiation acts as a mutagenic agent?
A 6–7 **B** 8–9 **C** 12–13 **D** 16–17

Items **11** and **12** refer to the accompanying table which shows the concentration of a non-biodegradable pesticide residue in the tissues of the organisms in a food chain and the water in their ecosystem.

	concentration of pesticide (ppm)
water	0.00005
plankton	0.04
herbivorous fish	0.23
carnivorous fish	2.07
fish-eating bird	6.00

11 The concentration of pesticide increased by a factor of nine times between
A herbivorous fish and carnivorous fish.
B carnivorous fish and fish-eating bird.
C plankton and herbivorous fish.
D water and plankton.

12 The concentration of pesticide in the fish-eating bird is greater than that in the water by a factor of
A 1.2×10^3. **B** 1.2×10^4. **C** 1.2×10^5. **D** 1.2×10^6.

Items **13** and **14** refer to the following graph which shows the results from a survey done on the number of lichen species growing along a 20 km transect from the centre of a city out to a country area.

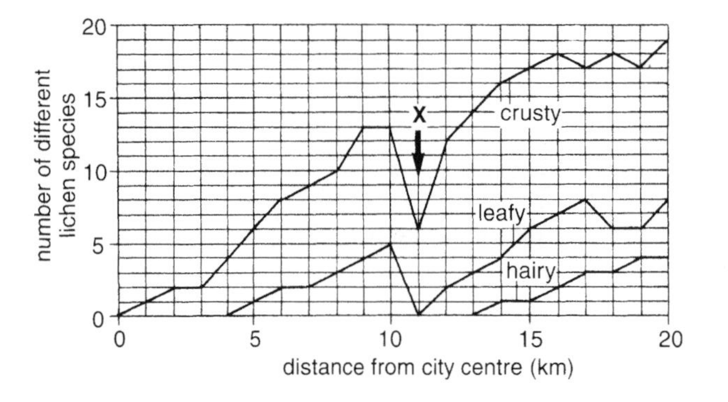

13 The dip in the graph at arrow X indicates
A an area of especially clean air.
B a local increase in sulphur dioxide concentration of air.
C an area lacking both hairy and crusty lichens.
D a lower level of atmospheric pollution compared with the country area.

14 Twenty-eight different species of lichen were recorded at one of the sites. The distance (in km) of this site from the city centre was
A 16. **B** 17. **C** 18. **D** 19.

15 The graph below follows the progress of an algal bloom and two associated environmental factors over the course of a year in the North Atlantic Ocean.

Which line in the table gives the correct identity of W, X and Y?

	relative number of algae	*relative light intensity*	*relative concentration of nitrate*
A	W	Y	X
B	X	W	Y
C	W	X	Y
D	X	Y	W

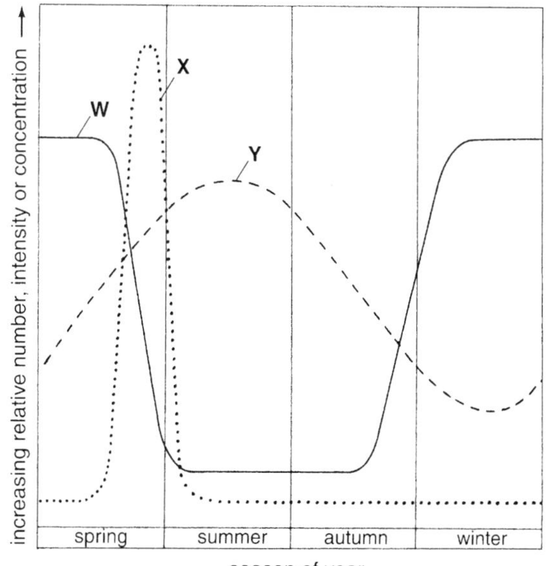

36 Succession in plant communities

Match the terms in list X with their descriptions in list Y.

list X		list Y	
1	biomass	**a**	dead organic matter that is added to soil during plant succession
2	climax community	**b**	unidirectional progression from pioneer to climax community by colonisation of an area that has been colonised before
3	climate and soil type	**c**	unidirectional progression from pioneer to climax community beginning on a barren uncolonised area
4	habitat modification	**d**	the total weight of living matter present in a community
5	humus	**e**	environmental factors affecting direction taken by succession
6	pioneer community	**f**	unidirectional long term change to an environment caused by plant succession
7	primary succession	**g**	first colonisers of an environment about to be modified by plant succession
8	secondary succession	**h**	final colonisers of an environment that has been modified by plant succession

Choose the ONE correct answer to each of the following multiple choice questions.

1 The following statements refer to the process of ecological succession. Which one is INCORRECT?

A A new group of plant species achieves dominance and ousts the old ones.
B The height and biomass of the vegetation increases as the process proceeds.
C Each group of species modifies the habitat making it more favourable for some other species.
D The number and variety of animal species decreases as the process proceeds.

2 Primary succession would occur on

A an abandoned field.
B a disused railway line.
C a rock surface bared by glaciation.
D a garden neglected by its owner.

3 Which of the following is NOT a characteristic of a climax community?

A It is the final product of long term change within the community.
B It is an immature community still to reach dynamic equilibrium with the environment.
C It is self-perpetuating and not replaced by another community.
D It is a stable community which does not show directional change.

4 As a newly bared mountain scree underwent succession, it became populated by the plant communities given in the list below.

1 shrubs **2** mosses **3** small trees
4 lichens **5** grasses

These communities would have appeared in the order

A 2, 4, 5, 1, 3. **B** 2, 5, 4, 3, 1. **C** 4, 2, 5, 1, 3. **D** 4, 5, 2, 3, 1.

5 The following table compares two types of community. Which line is INACCURATE?

	characteristic	*climax community*	*pioneer community*
A	diversity of species	high	low
B	life span of community members	long	short
C	food webs supported	complex	simple
D	nutrient supply in soil	low	high

6 The natural climax community in lowland Scotland is
A forest. **B** grassland. **C** moorland. **D** peat bog.

7 The natural climax community typical of a geographical region is determined by BOTH
A prevailing climate and soil type.
B soil type and land reclamation.
C land reclamation and agricultural practices.
D agricultural practices and prevailing climate.

8 In most parts of Britain, the vegetation is prevented from reverting to the natural climax community by
A human activities.
B prevailing climate.
C underlying soil type.
D competition between species.

Items 9 and 10 refer to the following information.

A freshwater lake may undergo plant succession if it has an inflow stream which carries silt as shown in the diagrams (see next page).

9 The sequence in which this series of events would occur is
A R, S, U, Q, P, T. **B** R, S, Q, U, P, T.
C R, Q, S, U, P, T. **D** R, S, Q, P, U, T.

10 Which of the following statements is NOT correct?
A Tussock-forming sedges make the environment more favourable for alder trees.
B Aquatic plants make the environment more favourable for reeds.
C Alder trees make the environment more favourable for oak trees.
D Sedges make the environment more favourable for reeds.

alder trees grow in marshy soil and increase its nitrogen content

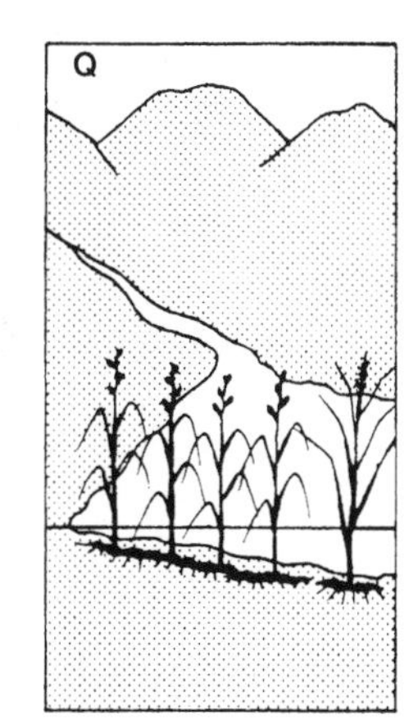

sedges thrive on accumulated silt

aquatic plants interfere with free flow of water making it deposit silt

reeds with underground rhizomes grow in silt and gather more sediment

oak forest becomes established on fertile soil

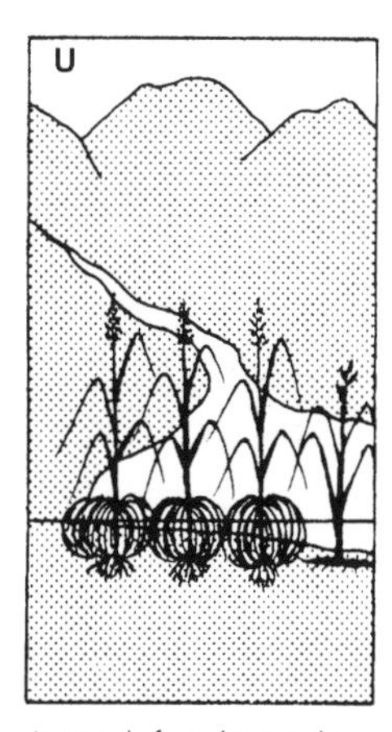

tussock-forming sedges form mounds of marshy soil made of silt and humus

Specimen Examination Paper 1

1 Which of the cells below would possess the greatest number of mitochondria per unit volume of cell?

A guard cell **B** motile sperm
C red blood cell **D** palisade mesophyll

2 The graph below shows the relationship between ion uptake by dandelion root cells and distance from root apex.

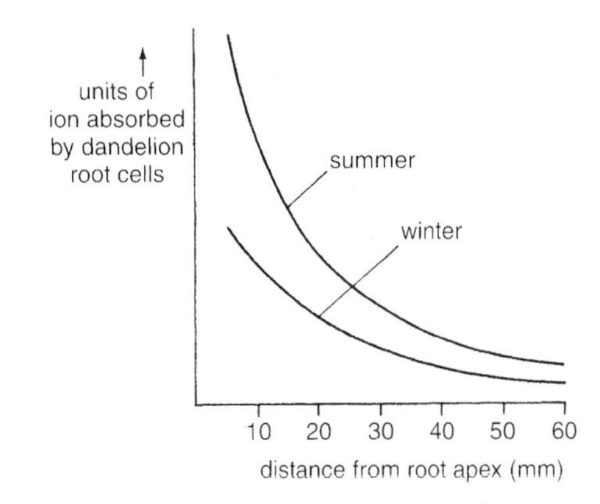

From these results it can be correctly concluded that in dandelion roots, most ion uptake occurs in

A young newly formed cells in summer.
B young newly formed cells in winter.
C older differentiated cells in summer.
D older differentiated cells in winter.

3 In an experiment, groups of potato discs were weighed and then each group was immersed in one of a series of sucrose solutions. After two hours each group was reweighed and its percentage gain or loss in weight was calculated.

The following graph shows the results plotted as points.

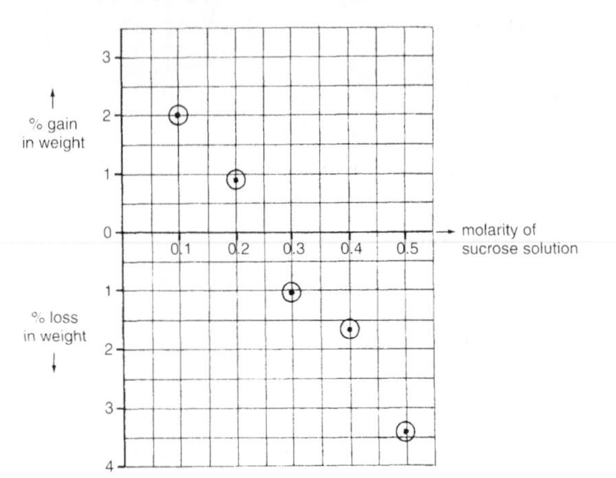

From these results it can be concluded that the water concentration of potato cell sap is approximately equivalent to that of a sucrose solution of molarity

A 0.10. **B** 0.25.
C 0.35. **D** 0.50.

4 The diagram below shows a simplified version of the Calvin cycle.

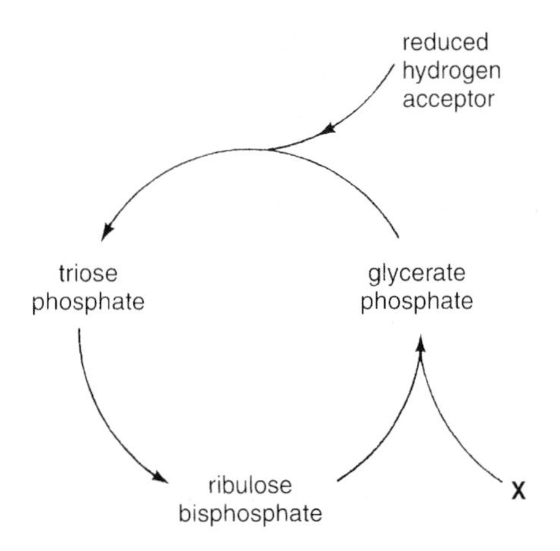

The correct identity of substance X is
A water. **B** glucose. **C** oxygen.
D carbon dioxide.

5 If a DNA molecule contains 5000 base molecules of which 22% are adenine, then the number of guanine molecules present is
A 1200. **B** 1400. **C** 2400.
D 2800.

6 The following list gives four stages involved in protein synthesis
1 tRNA and mRNA meet in a ribosome
2 region of DNA opens up exposing bases
3 mRNA is translated into a polypeptide chain
4 mRNA is transcribed using free nucleotides

These would occur in the sequence:
A 2, 4, 1, 3. **B** 2, 1, 4, 3.
C 4, 2, 1, 3 **D** 4, 1, 3, 2.

7 All viral particles possess
A a nucleic acid molecule and a hollow tail.
B a hollow tail and an enzyme molecule.
C an enzyme molecule and a protective coat.
D a protective coat and a nucleic acid molecule.

8 Which of the following does NOT occur during the first meiotic division?
A complete separation of chromatids
B crossing over between chromatids
C pairing of homologous chromosomes
D reduction of chromosome number

9 In a certain type of guinea pig, long hair (L) is dominant to short (l) and straight hair (S) is dominant to wavy (s). It is found that the cross LlSs × llss results in the ratio:

1 long, straight: 1 long, wavy: 1 short, straight: 1 short, wavy.

Which of the following crosses would also produce this result?
A LLSs × llss **B** Llss × llSS
C Llss × llSs **D** LlSs × LlSs

10 Red-green colour blindness is a sex-linked trait in humans. A colour-blind man marries a woman who is heterozygous for this condition. What proportion of their sons are likely to be colour-blind?
A 25% **B** 50% **C** 75% **D** 100%

11 What name is given to the type of gene mutation illustrated in the following diagram?
A inversion **B** deletion
C insertion **D** substitution

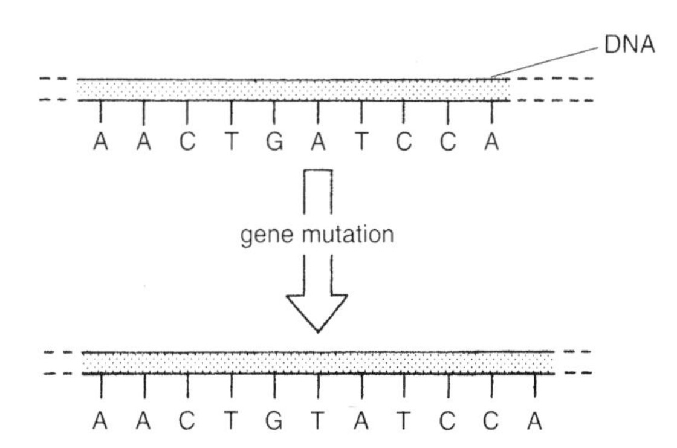

12 The chromosome shown in the diagram below became broken at the points indicated by arrows and the genes between these points became inverted.

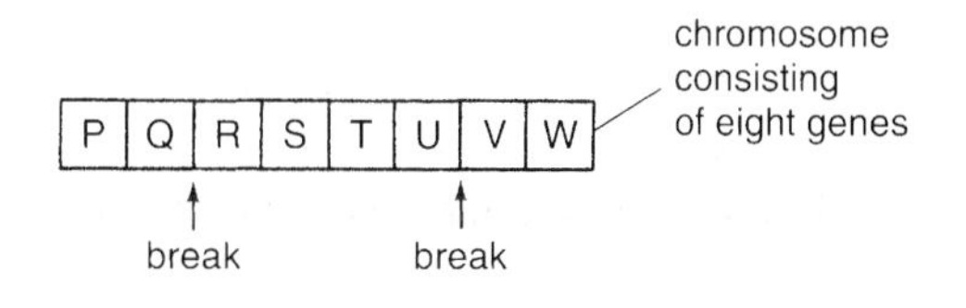

The resulting order of the genes was
A PQUTSRVW. **B** WVUTSRQP.
C PQTURSVW. **D** VWUTSRPQ.

13 Which of the following BOTH take place when a population suddenly finds itself in a favourable environment lacking competition and predation?

	selection pressure	*population size*
A	increases	increases
B	decreases	increases
C	increases	decreases
D	decreases	decreases

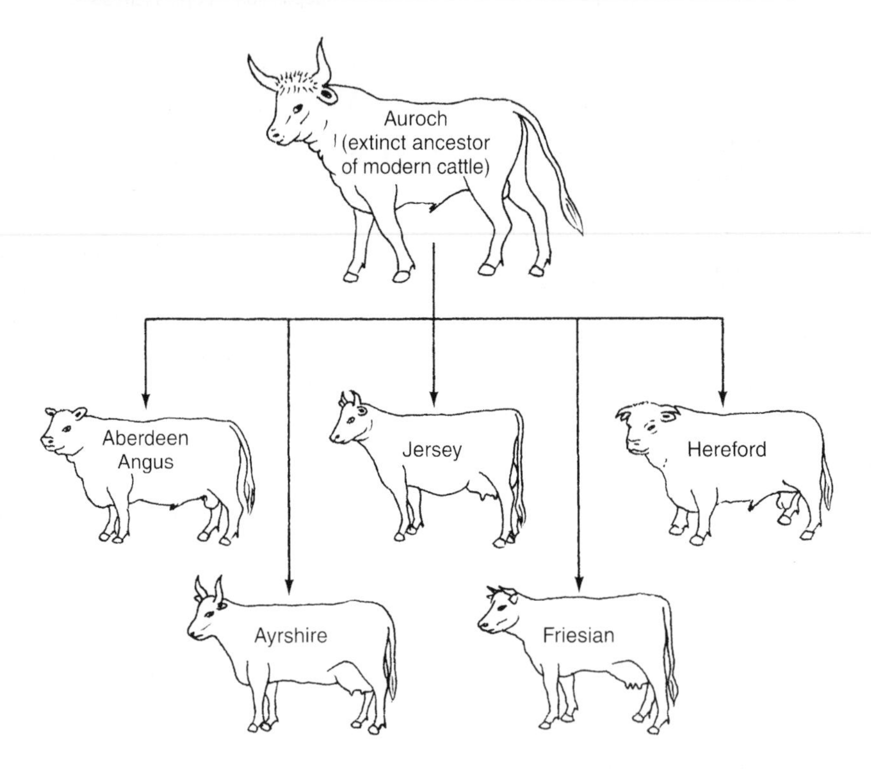

14 The evolution of various breeds of cattle is shown in the diagram above.
This has occurred mainly by

A natural selection of existing alleles in the gene pool.
B natural selection of newly mutated genes.
C artificial selection of existing alleles in the gene pool.
D artificial selection of newly mutated alleles.

15 Continuous inbreeding in an isolated population leads to

A increase in phenotypic variation.
B decrease in population size.
C increase in mutation rate.
D decrease in genetic variation.

16 The following diagram shows a piece of human DNA ready to be sealed into a bacterial plasmid.

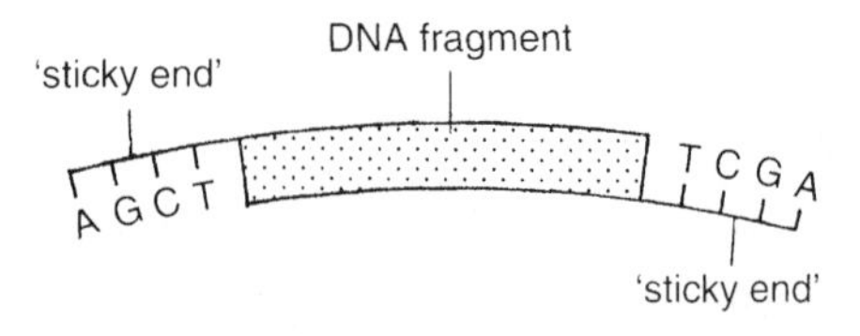

Which of the diagrams at the top of page 205 represents a suitable plasmid?

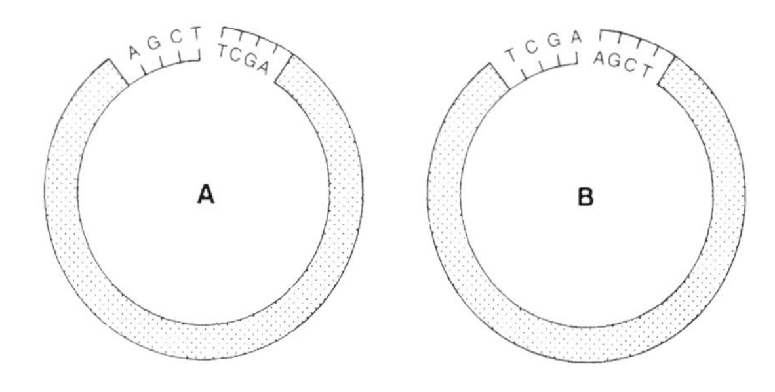

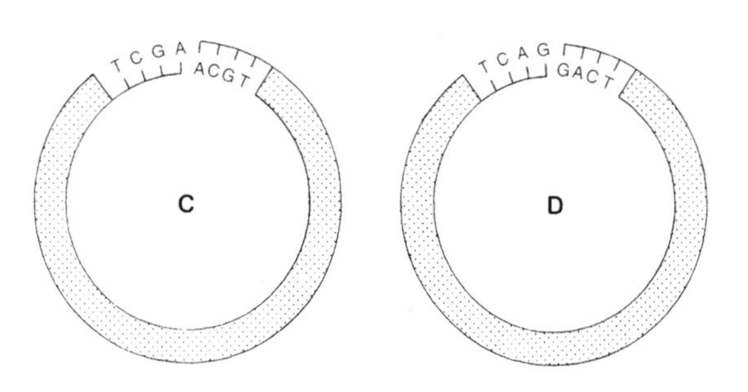

17 In the table below, * indicates that the bird in the left column dominated the other bird by pecking it.

↓ *bird against bird* →	4	3	2	1
1	*	*		
2	*	*		*
3				
4		*		

The pecking order of the birds is

A 1, 2, 3, 4. **B** 3, 4, 1, 2.
C 2, 1, 4, 3. **D** 4, 3, 2, 1.

18 It is best to transplant young seedlings on a day when the air is
A still, warm and dry.
B windy, warm and humid.
C still, cool and humid.
D windy, cool and dry.

19 The diagram that follows shows four of the stages that occur during the differentiation of phloem tissue in a plant.

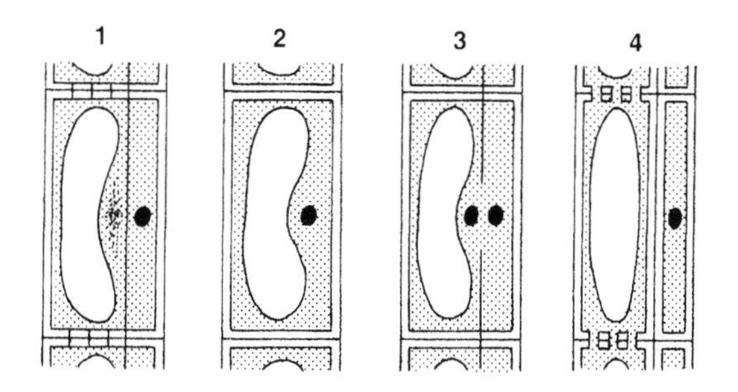

These occur in the order

A 2, 3, 1, 4. **B** 4, 1, 3, 2.
C 2, 3, 4, 1. **D** 3, 2, 1, 4.

Items **20, 21** and **22** refer to the following diagram which shows a possible arrangement of the genes involved in the induction of the enzyme β-galactosidase in the bacterium *Escherichia coli.*

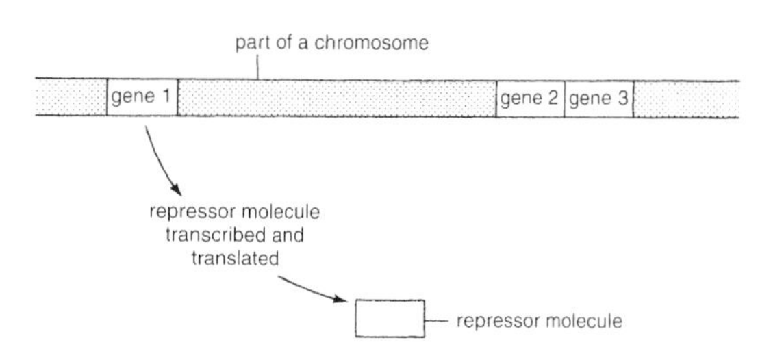

20 In the absence of lactose, the repressor molecule combines with gene 2 and, as a result, gene 3 remains 'switched off'. The correct identity of the three genes is

	Operator gene	*Structural gene*	*Regulator gene*
A	2	1	3
B	1	3	2
C	3	2	1
D	2	3	1

21 In this system, the operon consists of
A gene 2 only. **B** gene 3 only.
C genes 2 and 3. **D** genes 1, 2 and 3.

22 Which of the following situations would arise if lactose became available to the cell?

	gene 1	*gene 2*	*gene 3*
A	–	–	+
B	–	+	–
C	+	–	+
D	+	+	+

(+ = 'switched on', – = 'switched off')

23 Compared with the normal plant in the following diagram, which of the others shows symptoms typical of nitrogen deficiency?

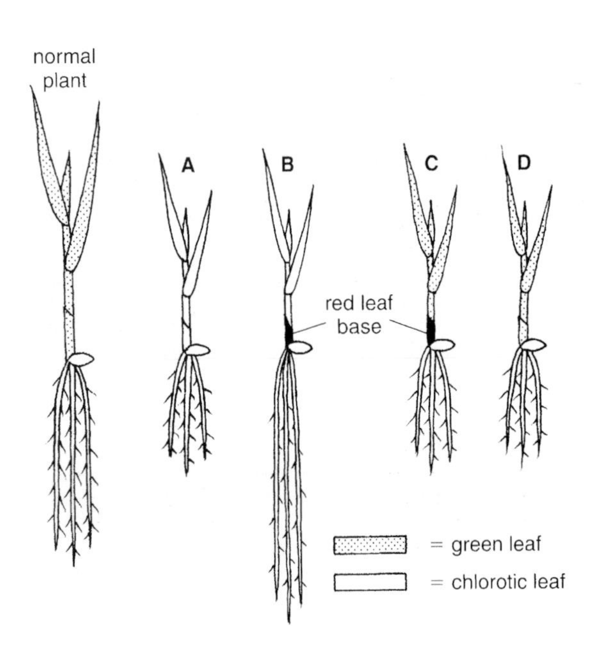

24 In the human body, which of the following does NOT occur in response to a sudden rise in environmental temperature?
A involuntary rhythmical contractions of skeletal muscles
B increase in blood circulation to body extremities
C increase in activity of sweat glands
D relaxation of erector muscles attached to hairs

25 Which of the following accurately represents the formula used to calculate the overall increase or decrease in size of a population? (br = birth rate, dr = death rate, er = emigration rate and ir = immigration rate)
A br – ir + er – dr
B br – dr + er – ir
C br – ir + dr – er
D br – dr + ir – er

26 Which of the following populations monitored by scientists acts as a vector of serious disease?
A phytoplankton **B** lichen
C stonefly **D** brown rat

27 The following table compares two types of community. Which line is CORRECT?

	characteristic	*pioneer community*	*climax community*
A	growth rate	slow	rapid
B	biomass	small	large
C	relative number of seeds	few	many
D	height of vegetation	high	low

28 An individual organism's development depends on the interaction between its
A genotype and environment.
B environment and phenotype.
C phenotype and chromosomes.
D chromosomes and genotype.

Items 29 and 30 refer to the table of results at the bottom of the page from a survey of breeding success in golden eagles.

29 The data support the hypothesis that an inverse relationship exists between thickness of shell and
A number of nests examined.
B average mass of egg contents.
C percentage of nests suffering theft.
D average concentration of dieldrin in egg.

30 In which year were five nests found to contain broken eggs?
A 3 **B** 4 **C** 5 **D** 6

year	*number of nests examined*	*percentage of nests with broken eggs*	*percentage of nests suffering theft of eggs*	*average mass of egg contents per nest (g)*	*average concentration of dieldrin in egg (ppm)*
1	17	5.90	5.90	127	0.03
2	20	10.00	10.00	109	0.19
3	15	13.30	20.00	143	0.53
4	16	18.80	6.23	115	2.11
5	18	27.80	11.10	104	6.63
6	19	31.60	15.80	132	7.15

Specimen Examination Paper 2

1 Which of the following acids is the final product of glycolysis?
A citric **B** lactic
C pyruvic **D** tricarboxylic

Items 2 and 3 refer to the accompanying diagram which shows ways in which molecules may move into or out of a respiring animal cell.

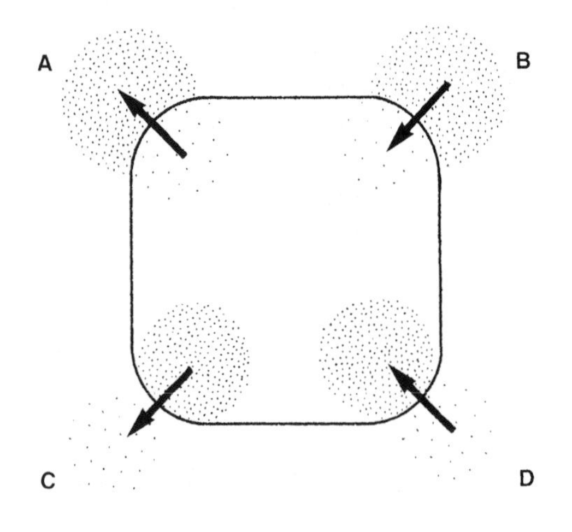

2 Which of these could be active transport of sodium ions out of the cell?

3 Which of these could be active uptake of potassium ions?

4 An area of grassland under study by scientists was found to receive 2×10^6 kJ of solar energy over a period of one year. The fate of the energy NOT used for photosynthesis was found to be as follows.

reflected by leaves	2.2×10^5 kJ
transmitted to the ground	8.0×10^4 kJ
converted to heat and lost	1.6×10^6 kJ

The percentage of light used annually for photosynthesis was
A 1. **B** 2. **C** 5. **D** 10.

5 Which of the following refers to the light-dependent stage of photosynthesis in a green plant cell?

	process	*location in chloroplast*
A	generation of ATP	stroma
B	fixation of carbon	grana
C	evolution of oxygen	stroma
D	reduction of hydrogen acceptor	grana

6 Successful replication of chromosomes does NOT require the presence of

A ribosomes. **B** a DNA template.
C nuclear enzymes. **D** adenosine triphosphate.

7 All viruses contain

A DNA. **B** RNA.
C DNA and RNA. **D** DNA or RNA.

8 A retrovirus differs from other types of virus in that it

A contains RNA not DNA.
B contains DNA not RNA.
C lacks reverse transcriptase.
D possesses reverse polymerase.

9 A person acquires long-lasting natural immunity to a particular antigen when she or he

A receives a small dose of vaccine by injection.
B is given antibodies produced by another mammal.
C responds to the antigen by producing antibodies.
D receives antibodies as a baby during suckling.

10 Syndactyly is a genetically inherited condition in humans involving the joining of two or more fingers by a web of skin and muscle. It is determined by the presence of a dominant allele.

The key to the symbols used in the family tree shown below is as follows:

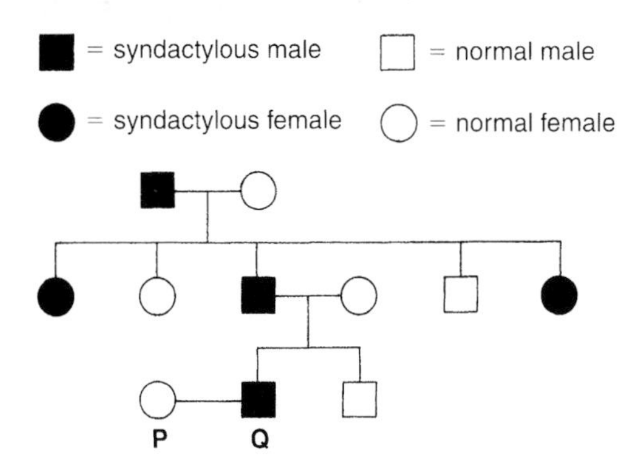

The chance of individuals P and Q producing a syndactylous child is

A 1 in 1. **B** 1 in 2.
C 1 in 3. **D** 1 in 4.

11 Three true-breeding female fruit flies with red eyes and black bodies were crossed with six true-breeding male pink-eyed, grey-bodied flies. All members of the F_1 generation were found to possess red eyes and grey bodies.

Three F_1 females were crossed with six F_1 males and produced an F_2 generation consisting of 115 red-eyed, grey: 37 pink-eyed, grey: 41 red-eyed, black: 12 pink-eyed, black.

The results indicate that inheritance of these two characteristics is determined by

A sex-linked genes.
B linked autosomal genes.
C genes on independently segregating chromosomes.
D one gene affecting both characteristics and showing incomplete dominance.

12 The following diagram shows two types of chromosome mutation.

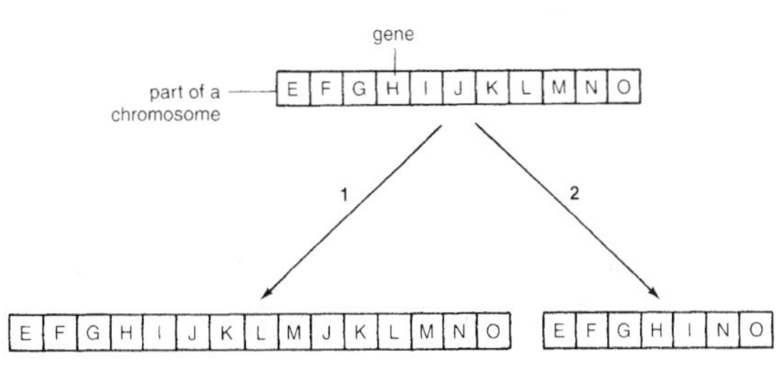

These are called

	1	2
A	duplication	deletion
B	duplication	substitution
C	inversion	deletion
D	inversion	substitution

13 A mutation can occur in any part of the body at any time. In fruit flies it is most likely to spread through a population if it is present in a

A wing cell. **B** eye cell.
C sperm cell. **D** salivary gland cell.

14 When first brought into clinical use, the antibiotic penicillin was found to be highly effective against the bacterium *Staphylococcus* which can cause many serious infections. However, within 10 years, many strains of *Staphylococcus* were found to be resistant to penicillin.

This situation arose because

A the bacterium's original gene pool already contained some alleles which determined resistance to penicillin.
B penicillin induced a mutation producing a resistance gene which was passed on to future generations of bacteria.
C all bacterial cells possess a few genes which are resistant to penicillin and enjoy a selective advantage.
D penicillin stimulated the development of resistance in some bacteria and these were later selected during infection.

15 Which of the following acts as a reproductive barrier during the process of speciation?

A river formed at the end of an ice age.
B failure of pollination mechanism.
C difference in altitude of habitats.
D alteration of pH by excessive use of fertilizer.

16 The chromosomes shown in the opposite diagram have been stained and therefore show characteristic banding patterns. Which chromosome is abnormal in that it has undergone duplication of part of its genetic material?

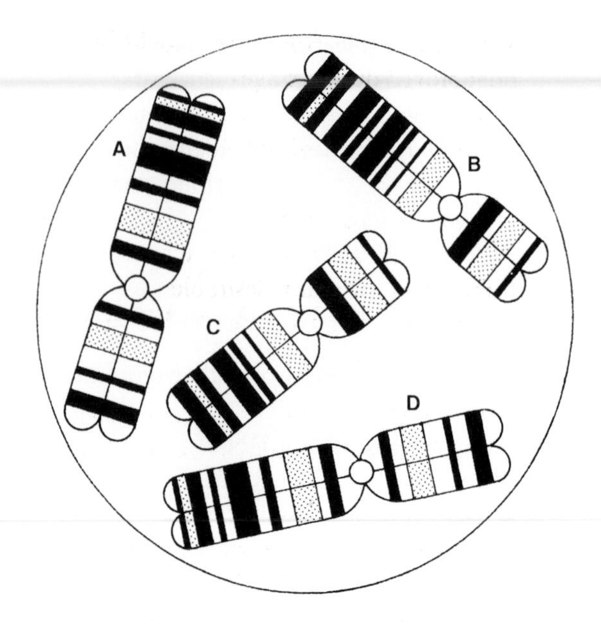

Items **17** and **18** refer to the following list of procedural steps employed during recombinant DNA technology.

1 host cell allowed to multiply
2 required DNA fragment cut out of appropriate chromosome
3 duplicate plasmids formed which express 'foreign' gene
4 plasmid extracted from bacterium and opened up
5 recombinant plasmid inserted into bacterial host cell
6 DNA fragment sealed into plasmid

17 The correct order in which these steps would be carried out is

A 2, 4, 6, 5, 1, 3. **B** 4, 6, 2, 5, 3, 1.
C 2, 4, 5, 6, 3, 1. **D** 4, 6, 2, 5, 1, 3.

18 The enzyme endonuclease would be employed during stages
A 2 and 6. **B** 2 and 4.
C 4 and 5. **D** 4 and 6.

19 P_1 in the cross shown below represents a cultivated variety of plant which contains the genes for many desirable characteristics. However, P_1 is susceptible to a particular virus.

P_2 is a wild variety of the same species which possesses the gene for resistance to the virus.

Following the first cross, 50% of the genetic material (including the resistance gene) received by P_3 comes from the wild parent. In order to dilute this unwanted wild genetic contribution but retain the resistance gene, a series of crosses against P_1 is carried out. The first is shown in the diagram and results in the formation of P_4.

How many more crosses will need to be carried out to reduce the wild genetic material inherited by offspring to less than 5%?

A 1 **B** 2 **C** 3 **D** 4

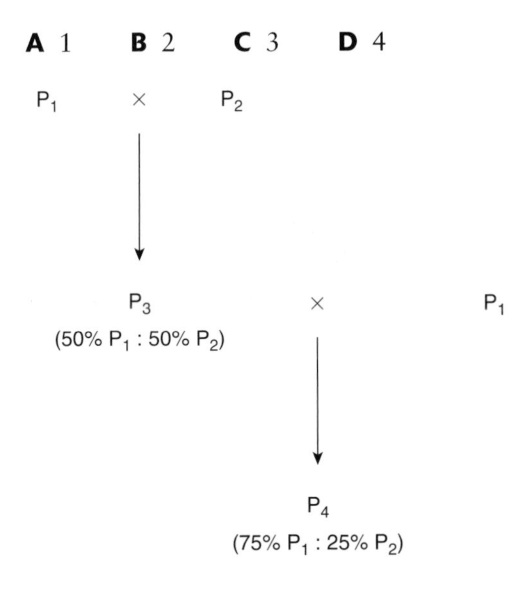

20 The gills of a freshwater bony fish
A gain water by osmosis and absorb salts.
B lose water by osmosis and absorb salts.
C gain water by osmosis and excrete salts.
D lose water by osmosis and excrete salts.

21 At the compensation point in a green plant
A photosynthesis comes to a halt.
B rate of respiration exceeds rate of photosynthesis.
C evolution of oxygen exceeds uptake of carbon dioxide.
D rate of synthesis of glucose equals rate at which it is used up.

22 The accompanying diagram shows a sigmoid growth curve for a population of yeast cells. The period of steady rapid growth occurred between
A the start and day 5.
B day 5 and day 10.
C day 10 and day 15.
D day 15 and day 20.

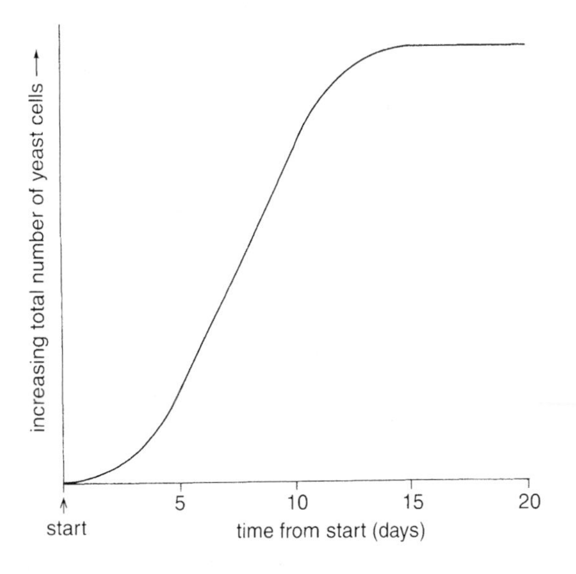

23 The diagram below shows an experiment set up to investigate abscission of leaf stalks.

Following the treatment shown in the diagram, abscission of the leaf stalk would occur in

A plant 1 only. **B** plants 1 and 2.
C plants 2 and 3. **D** plant 3 only.

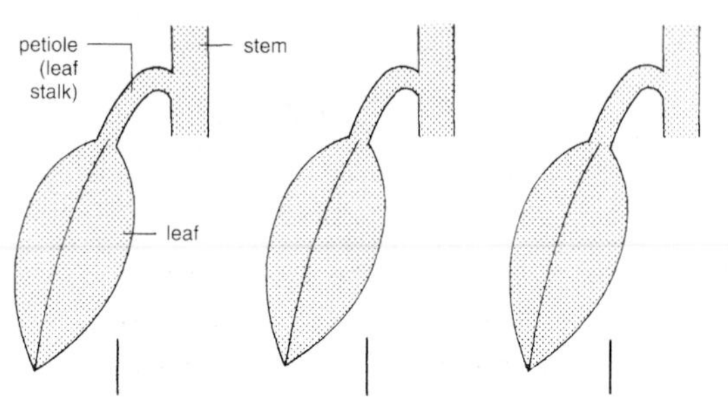

leaves cut off leaving petioles (leaf stalks) intact and then treated as follows

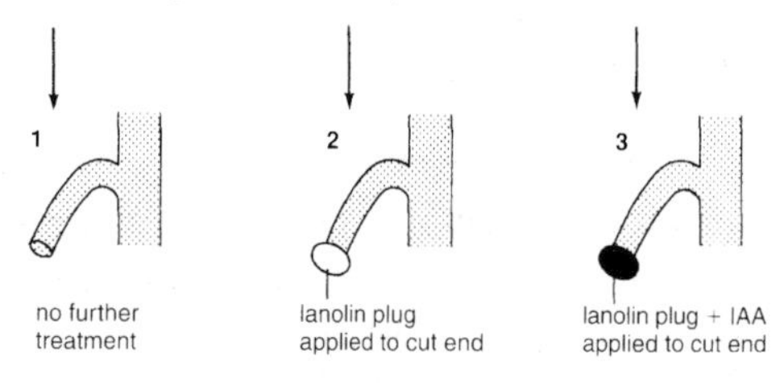

24 The three numbered structures in the diagram below represent endocrine glands.

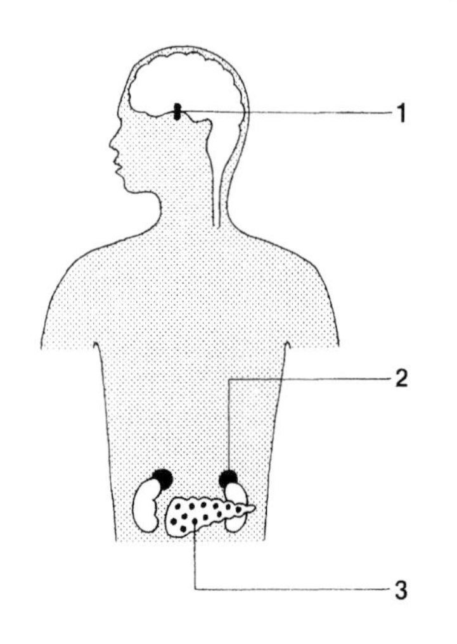

Which of the following combinations correctly matches these glands with the hormones that they produce?

	endocrine gland		
	1	2	3
A	growth hormone	adrenaline	ADH
B	growth hormone	ADH	insulin
C	ADH	adrenaline	glucagon
D	ADH	glucagon	insulin

(ADH = anti-diuretic hormone)

25 An etiolated plant is characterised by possessing BOTH

A expanded yellow leaves and long strong internodes.
B curled yellow leaves and long weak internodes.
C expanded green leaves and short strong internodes.
D curled green leaves and short weak internodes.

26 The table below refers to four species of whale.

species	*estimated number before commercial whaling*	*estimated number today*
Humpback	100 000	9000
Fin	450 000	70 000
Sei	200 000	28 000
Right	50 000	3000

Based on these estimates, which species of whale has undergone the greatest relative decrease in numbers?

A Humpback **B** Fin **C** Sei **D** Right

27 Which of the following is NOT a source of heritable variation?

A independent assortment of chromosomes
B crossing over during meiosis
C gene mutations
D natural selection

28 Which of the following ecosystems would tend to remain most stable?

	relative state of ecosystem	*predator-prey relationships*
A	simple	only one prey species for each predator
B	complex	only one prey species for each predator
C	simple	many prey species for each predator
D	complex	many prey species for each predator

Items 29 and 30 refer to the accompanying diagram which shows a simplified version of the nitrogen cycle.

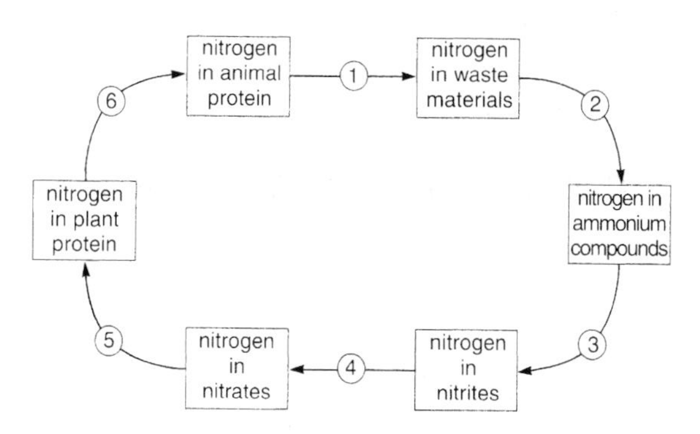

29 Effluent from adequately treated sewage when discharged into a river may lead to an algal bloom. Which arrow represents the stage in the cycle at which this growth would occur?

A 3 **B** 4 **C** 5 **D** 6

30 Untreated sewage may be discharged into a river and promote the growth of sewage fungus. At which arrow in the diagram would this process occur?

A 1 **B** 2 **C** 3 **D** 4

ITEM NUMBER

TEST NUMBER

	1	2	3	4	5	6	7	8	9	10	11	12	13	14	15	16	17	18	19	20
1																				
2																				
3																				
4																				
5																				
6																				
7																				
8																				
9																				
10																				
11																				
12																				
13																				
14																				
15																				
16																				
17																				
18																				
19																				
20																				
21																				
22																				
23																				
24																				
25																				
26																				
27																				
28																				
29																				
30																				
31																				
32																				
33																				
34																				
35																				
36																				

SPECIMEN EXAMINATION PAPER 1

1	2	3	4	5	6	7	8	9	10	11	12	13	14	15
16	17	18	19	20	21	22	23	24	25	26	27	28	29	30

SPECIMEN EXAMINATION PAPER 2

1	2	3	4	5	6	7	8	9	10	11	12	13	14	15
16	17	18	19	20	21	22	23	24	25	26	27	28	29	30

ITEM NUMBER

TEST NUMBER

	1	2	3	4	5	6	7	8	9	10	11	12	13	14	15	16	17	18	19	20
1	C	D	D	A	B	C	C	A	B	B	A	B	B	D	D	D	B	A	C	A
2	B	C	A	A	D	A	D	B	B	B	C	D	D	A	B	A	C	D	C	C
3	B	A	D	C	A	B	B	C	D	C										
4	A	C	D	B	C	D	B	D	C	A	B	B	A	D	A	C	B	D	A	C
5	A	D	A	B	C	A	B	D	D	C	D	B	C	B	C					
6	C	A	D	A	C	D	D	B	B	A	B	C	D	A	B					
7	C	D	B	D	D	A	C	B	A	B	A	C	D	C	B					
8	B	A	B	D	A	C	A	C	C	B	C	D	B	D	A	C	A	D	D	B
9	C	D	A	A	C	B	C	B	D	A										
10	B	A	B	A	D	C	A	B	C	D	D	C	A	C	B					
11	C	A	A	C	B	C	B	D	D	D	C	D	B	A	B	A	B	C	C	D
12	D	C	B	C	A	D	D	B	D	B	A	A	C	C	A					
13	D	D	A	B	B	A	C	C	C	B										
14	A	D	A	B	C	D	B	B	D	C	D	C	B	A	C					
15	B	B	D	A	C	C	B	C	A	D	A	D	B	C	D					
16	D	B	A	C	C	A	D	D	B	A	D	A	B	C	B					
17	C	D	C	C	D	B	A	B	A	C	D	A	C	B	A	B	B	D	B	A
18	D	D	C	C	B	A	C	D	B	D	D	A	C	A	A	B	A	B	B	C
19	C	B	A	A	D	C	D	B	C	B										
20	C	D	B	A	D	A	B	C	B	A										
21	B	C	B	C	C	C	D	D	A	B	A	D	B	B	A	C	A	A	D	D
22	B	A	C	D	D	A	C	C	D	B	B	A	B	C	A					
23	A	B	A	B	D	D	C	A	B	D	C	D	A	B	C	C	B	D	A	C
24	C	D	D	A	A	C	B	A	D	A	B	C	B	D	C					
25	A	D	B	A	D	B	C	C	B	D										
26	B	B	D	A	A	B	C	B	D	B	C	C	A	A	D	C	D	A	B	D
27	B	A	C	A	C	D	D	A	D	B	C	B	B	D	D	A	C	A	C	B
28	B	C	A	D	B	A	C	C	D	A	B	C	D	A	D					
29	A	C	D	A	D	A	D	B	B	C	C	D	D	B	C	A	B	A	C	B
30	B	C	C	D	C	C	D	D	A	A	A	B	B	C	A	B	D	B	D	A
31	A	B	B	C	D	A	C	C	A	D	B	D	C	A	D					
32	D	B	C	A	B	D	A	B	D	C										
33	B	A	A	C	A	D	A	C	C	B	B	B	C	C	D	D	B	A	D	B
34	B	C	B	D	D	C	A	A	D	A	A	B	B	C	D					
35	D	C	A	C	A	C	B	B	D	C	A	C	B	B	D					
36	D	C	B	C	D	A	A	A	B	D										

SPECIMEN EXAMINATION PAPER 1

1	2	3	4	5	6	7	8	9	10	11	12	13	14	15
B	A	B	D	B	A	D	A	C	B	C	A	B	C	D
16	17	18	19	20	21	22	23	24	25	26	27	28	29	30
B	C	C	A	D	C	D	B	A	D	D	B	A	D	C

SPECIMEN EXAMINATION PAPER 2

1	2	3	4	5	6	7	8	9	10	11	12	13	14	15
C	A	D	C	D	A	D	A	C	B	C	A	C	A	B
16	17	18	19	20	21	22	23	24	25	26	27	28	29	30
B	A	B	C	A	D	B	B	C	B	D	D	D	C	B